Gernot Hickethier

Communication Structures in the Design Phase of Lean Project Delivery

Karlsruher Reihe Technologie und Management im Baubetrieb
Karlsruher Institut für Technologie (KIT)
Institut für Technologie und Management im Baubetrieb

Prof. Dr.-Ing. Dipl.-Kfm. Shervin Haghsheno (Hrsg.)
Prof. Dr.-Ing. Sascha Gentes (Hrsg.)

Heft 71

Das Institut für Technologie und Management im Baubetrieb (TMB) befasst sich in Forschung und Lehre mit dem gesamten Bereich des Baubetriebs von der Maschinen- und Verfahrenstechnik bis hin zum Management der Projekte, Facilities und Unternehmen.
Weitere Informationen und Kontakte unter www.tmb.kit.edu

Eine Übersicht der Forschungsberichte finden Sie am Ende des Buches.

Communication Structures in the Design Phase of Lean Project Delivery

by
Gernot Hickethier

KIT Scientific Publishing

Karlsruher Institut für Technologie
Institut für Technologie und Management im Baubetrieb

Communication Structures in the Design Phase of Lean Project Delivery

Zur Erlangung des akademischen Grades eines Doktor-Ingenieurs von der KIT-Fakultät für Bauingenieur-, Geo- und Umweltwissenschaften des Karlsruher Instituts für Technologie (KIT) genehmigte Dissertation

von Dipl.-Wi.-Ing. Gernot Hickethier aus Marburg

Tag der mündlichen Prüfung: 21. Juli 2015
Referent: Prof. Dr.-Ing. Fritz Gehbauer, M.S.
Korreferentin: Prof. Iris D. Tommelein, Ph.D.

Impressum

KIT Scientific Publishing

Karlsruher Institut für Technologie (KIT)
KIT Scientific Publishing
Straße am Forum 2
D-76131 Karlsruhe

KIT Scientific Publishing is a registered trademark
of Karlsruhe Institute of Technology.
Reprint using the book cover is not allowed.

www.ksp.kit.edu

Print on Demand 2019 – Gedruckt auf FSC-zertifiziertem Papier

ISSN 2363-8222
ISBN 978-3-7315-0802-1
DOI 10.5445/KSP/1000083263

Foreword

One of the characteristics of Lean Project Delivery is to establish improved communication between the stakeholders in the process with the aim to improve co-operation and to reveal improvements for the product even if interests diverge at the beginning. Special forms of contract and incentive schemes have been developed for this purpose. Control mechanisms exist, e.g. the measurement of the reliability of commitments. All this is feasible only if a good system and practice of communication is in place. Whereas factors as reliability, target cost and value, and workflow can be measured Gernot Hickethier reveals the fact that models to monitor planned and actual communication do not exist. His aim is to present a method of doing this and to propose a procedure to draw conclusions from it for improvement of the communication structure.

In the introduction the research questions and relevant research method are outlined. A substantial literature review (chapter 2) leads to the potentials of using Forced Directed Graphs and Matrix-based Models (DSM, DMM) for the purpose of the research and reveals the research gap described in chapter 3. This leads to a Social Network Analysis of an ongoing complex design project with some preliminary findings and recommendations (chapter 4). The same project is used for one of the case studies presented in chapter 6. Chapter 5 presents a Delta-Analysis for comparing "planned" and "as-is" communication subsequently applied in two case studies (chapter 6). To find the "delta" two indicators are proposed: The existing BIM development process for the planned communication and the recorded BIM clashes for the actual communication. Chapter 7 presents the conclusions.

A very good insight is given to the importance of communication in complex design projects framed in Integrated Project Delivery (IPD). A remarkable effort is made to develop models for planned and actual communication and to guide the project team -through workshops- into discussion of possible and necessary amendments as conclusion from the Delta Analysis.

The importance of communication in complex design with a multitude of designers and design tasks is out of question. The author introduces the Design Structure Matrix and Forced Directed Graphs to visualize and measure communication processes and to facilitate improvement considerations through a Delta Analysis. This is a contribution to the science of Lean Management. The modeling of actual information flow using the BIM data is unique. All this is based on observations on projects and the conclusions contain relevant recommendations for practical applications. The result is a new procedure with model. Finally, Gernot Hickethier has guided a design team to use the model and to interpret the findings.

Prof. Dr.-Ing. Fritz Gehbauer, M.S.

Acknowledgements

This work mostly results from my occupation as a researcher at the Institute for Technology and Management in Construction. First, I would like to thank my doctoral advisor, Fritz Gehbauer, for his continuing support, for granting the scientific freedom to try out new ideas, and for the invaluable experiences I gained during implementation of Lean methods in the industry. I am also grateful to my second advisor, Iris Tommelein, for the opportunity to pursue part of my research in the United States, for introducing me to the Lean Construction research community in the Bay Area, and for her valuable reviews and ideas, which helped in developing this dissertation to its current state. I am also thankful to Petra von Both and Albert Albers who served as members of my dissertation committee and who provided valuable perspectives from the fields of architecture and mechanical engineering.

Special thanks go to my research collaborators in the industry who helped make this dissertation possible: Damon Chandler, Michelle Hofmann, Volkmar Hovestadt, Atul Khanzode, Baris Lostuvali, Paul Reiser, Gerolf Sonntag, Andy Sparapani, and David Thomack.

An important thank you goes to Vera Hickethier and Alexander Lange who helped me proofread this work. I would like to thank my colleagues for intellectual stimulation, inspiration, and laughter: Tobias Bregenhorn, Michael Denzer, Shervin Haghsheno, Kim Kirchbach, Daniel Knecht, Ahlam Mohamad, Ana Schilling Miguel, Heiner Schlick, Harald Schneider, Ahmed Stifi, and Annett Schöttle at Karlsruhe Institute of Technology, and Isabel Alarcon, Paz Arroyo, Glenn Ballard, Stéphane Denerolle, Lynn Hiel, Hyun Woo Lee, Philip Lorenzo, Corinne Scown, and Patricia Tilman at University of California Berkeley.

I would also like to thank the Fulbright Program and the Karlsruhe House of Young Scientists for supporting my research stay in the United States.

Lastly and mostly, I am deeply grateful to my wife and my family for supporting me and believing in me. Their patience and motivation, even during stressful times, gave me the energy to complete this dissertation.

Stuttgart, June 2016 Gernot Hickethier

Summary

The thesis at hand focuses on communication in the detailed design phase of complex projects. Communication plays an important role in integrating project partners, including lateral communication that facilitates coordinating the project team. Communication within the project team is a prerequisite for reflecting tasks completed, and thus a prerequisite for continuous improvement.

Proponents of Lean Construction have advocated for project integration in theory and practice. For example, relational project contracts, developed in the United States and Australia, foster collaboration through aligning project partners' interests. Also, increased use of project-wide IT systems, for example, Building Information Modeling (BIM), demands stronger integration of processes between project partners.

The goal of this thesis is to develop a method that facilitates continuous improvement of communication structures in projects. First, the literature review examines existing approaches for modeling and improving communication structures. Next, the author derives requirements for the newly developed method and its application. A case study shows that these requirements are attainable. The author then presents the method for continuous improvement of communication structures and applies it successfully in two case studies.

The method for improvement of communication structures is based on the principles of Lean Management and applies delta-analysis between two communication models: a prescriptive model of planned communication and a descriptive model of actual communication. Delta-analysis, which utilizes methods from the field of structural complexity management and social network analysis, identifies differences between planned and actual communication. Force-directed graphs visualize results of the analysis and, in turn, facilitate the identification of opportunities for improvement.

This thesis consists of two separate but related parts. Part one tests whether the requirements "integration" and "flexibility" exist in integrated project organizations. Integrated Project Delivery (IPD) is a relational contract that aligns the project partners' particular interests to optimize the project as a whole. The author models the project organization of the Van Ness and Geary Campus (VNGC) Hospital Project as a social network based on a survey of the project team. A set of metrics enables analysis of the social networks and shows that the requirements "integration" and "flexibility" exist in the VNGC project organization. Regarding integration, the case study shows that designers and builders interact closely in cluster groups. Regarding flexibility, the case study shows that some people have central positions in the social network by taking on a coordinating role, even though it is not part of their job description.

Part two of this thesis applies the method for continuous improvement of communication structures in two case studies. Application of the method identifies opportunities for improvement in both case studies. The identified root-causes for misalignments between planned and actual communication originate in integrative mechanisms, planned processes, and project environment. The results of part two highlight the need to regard projects as open systems which interact with their environment.

This thesis presents a Social Network Analysis (SNA)-based approach for examining communication structures in projects. Application of the approach provides evidence of an integrated and flexible project organization in the design phase of an IPD project. Application of the method for continuous improvement of communication structures shows the benefit of comparing models of planned and actual communication. The case studies show that BIM data can be used to model actual communication. This creates a new use case for BIM as a data source for modeling of information flow in design processes. This use of existing data for modeling communication networks achieves transparency and reduces modeling effort significantly, thereby lowering the barrier for application of the method.

Kurzfassung

Die vorliegende Dissertation untersucht die Kommunikationsstrukturen in der Genehmigungsplanung komplexer Bauprojekte. Kommunikation spielt in der Integration der verschiedenen Projektpartner eine wichtige Rolle, auch als laterale Kommunikation mit dem Ziel der bottom-up Koordination des Projektteams. Die Kommunikation innerhalb eines Projektteams ist aber auch die Voraussetzung zur Reflektion der eigenen Tätigkeit und somit notwendig für die erfolgreiche Durchführung kontinuierlicher Verbesserungsprozesse.

In Theorie und Praxis wurde bereits erkannt, dass eine verstärkte Integration der Projektpartner am Bau erforderlich ist. Beispielsweise entwickelten sich relationale Vertragsmodelle im englischsprachigen Raum. Diese Art von Verträgen fördert die Zusammenarbeit durch das Angleichen der verschiedenen Einzelinteressen. Zugleich fordert die verstärkte IT-Integration die Verzahnung der Projektpartner. An dieser Stelle ist die Anwendung von Building Information Modeling (BIM) zu nennen.

Ziel dieser Arbeit ist es, eine Methode zu entwickeln, welche die kontinuierliche Verbesserung von Kommunikationsstrukturen im integrierten Projektumfeld unterstützt. Um dieses Ziel zu erreichen, werden bestehende Ansätze zur Beschreibung und Verbesserung von Kommunikationsstrukturen untersucht. Anforderungen an eine zu entwickelnde Methode und für deren erfolgreiche Anwendung werden definiert, sowie die Erreichbarkeit der Anforderungen für die erfolgreiche Anwendung überprüft. Darauf aufbauend wird eine Methode zur Verbesserung von Kommunikationsstrukturen entwickelt und diese in Fallstudien getestet.

Die Methode zur Verbesserung von Kommunikationsstrukturen ist verwurzelt in den Prinzipien des Lean Management und basiert auf der Delta-Analyse zwischen geplanter und tatsächlich stattfindender

Kommunikation. Zur Ermöglichung der Delta-Analyse werden zwei Kommunikationsmodelle erstellt, ein präskriptives der geplanten Kommunikation sowie ein deskriptives der tatsächlichen Kommunikation. Die Delta-Analyse nutzt Methoden des strukturellen Komplexitätsmanagements und der sozialen Netzwerkanalyse zur Untersuchung der Unterschiede zwischen geplanten und tatsächlich stattfindenden Kommunikationsstrukturen. Kräftebasierte Graphen werden angewandt, um die Ergebnisse der Untersuchung in einem Workshop mit dem Planungsteam zu visualisieren und so Verbesserungspotenziale zu identifizieren.

Im Rahmen dieser Dissertation wurden zwei getrennte, jedoch inhaltlich verknüpfte Studien durchgeführt. Die erste Studie untersucht, ob die Anforderungen „Integration" und „Flexibilität" in Organisationen integriert abgewickelter Projekte gegeben sind. Die integrierte Projektabwicklung (engl. Integrated Project Delivery – IPD) ist ein relationales Vertragsmodel, welches die Einzelinteressen der Projektpartner angleicht, um die Optimierung des Gesamtprojekts zu stärken. Zur Untersuchung der Anforderungen wird auf Basis einer Umfrage ein soziales Netzwerkmodell der Projektorganisation des Van Ness and Geary Campus (VNGC) Krankenhausprojekts in San Francisco, USA, erstellt. Anhand von Metriken der sozialen Netzwerkanalyse wird gezeigt, dass die Anforderungen „Integration" und „Flexibilität" im VNGC Projekt vorhanden sind. Die Studie zeigt bezüglich der Existenz von Integration, dass Planer und Ausführende in Cluster-Gruppen eng zusammenarbeiten. Betreffend der Existenz von Flexibilität kann in der Studie festgestellt werden, dass einige Personen zentrale Stellen innerhalb der tatsächlichen Kommunikationsstruktur der Projektorganisation einnehmen, obwohl Koordination nicht Teil ihrer Aufgabe ist.

In der zweiten Studie wird die entwickelte Methode zur Verbesserung von Kommunikationsstrukturen in zwei Fallstudien angewendet. Die Anwendung identifiziert Verbesserungspotenziale innerhalb der Projektorganisationen. Die identifizierten Gründe für Abweichungen zwischen geplanten und tatsächlichen Kommunikationsstrukturen liegen in den integrativen Mechanismen, den geplanten Prozessen und der Umwelt der Projektorganisationen. Die Ergebnisse verdeutlichen die Notwendigkeit, Projekte als offene, mit ihrer Umwelt interagierende Systeme zu betrachten.

Diese Arbeit präsentiert eine Methodik, anhand derer die Eigenschaften von Kommunikationsstrukturen in Projekten überprüft werden können. Die Anwendung der Methodik zeigt, dass das untersuchte IPD-Projekt die geforderten Eigenschaften besitzt. Die im Rahmen dieser Arbeit entwickelte Methode zur Verbesserung von Kommunikationsstrukturen begründet den Nutzen des Vergleichs von präskriptiven und deskriptiven Kommunikationsmodellen anhand der vorgestellten Fallstudien. Deskriptive Kommunikation kann durch Anwendung von Indikatoren für Kommunikation modelliert werden, und Datenbank-Protokolle des IT-Werkzeugs BIM können hierzu genutzt werden. Diese Art der Datengewinnung stellt eine weitere mögliche Nutzung von BIM dar, durch welche mit geringem Aufwand die tatsächliche Kommunikation in Projektteams transparent gemacht werden kann. Die Nutzung existierender Daten zur Modellierung der Kommunikationsnetzwerke reduziert den Aufwand der Modellerstellung und vereinfacht die Anwendung der Methode zur kontinuierlichen Verbesserung von Kommunikationsstrukturen maßgeblich.

Table of Contents

Index of Figures

Index of Tables

Table of Abbreviations

3D	3-dimensional
A	Additional
ADePT	Analytical Design and Planning Technique
AEC	Architecture-Engineering-Construction
APT	Advanced Pneumatic Tubes, Inc.
Arch.	Architect
BAGS	Bagatelos Architectural Glass Systems, Inc.
BIM	Building Information Model, Building Information Modeling
CBA	Choosing by Advantages
CHH	Cathedral Hill Hospital
CPMC	California Pacific Medical Center
DA	Design Assist
DB	Design-Build
DBB	Design-Bid-Build
DE	Degenkolb Engineers
DFI	Design for Integration
DJ	D&J Tile Company
DMM	Domain Mapping Matrix
DSM	Design Structure Matrix
DWG	Drawing; A drawing file format native to AutoCAD
E	Expected
Electr.	Electrical
FS	Fuel Oil Systems
FSG	Functional Support Group

GC	General Contractor
GMP	Guaranteed Maximum Price
GST	General Systems Theory
HB	Herrero Boldt
HVAC	Heating, ventilation, and air conditioning
IFOA	Integrated Form of Agreement
IGLC	International Group for Lean Construction
IM	Integrative Mechanism
IP	Impact and Procedure
IPD	Integrated Project Delivery
IPDT	Integrated Project Delivery Team
IPT	Integrated Product Team
ISAT	International Seismic Application Technologies
ISEC	ISEC, Inc.
KHSS	KHS&S Contractors
LAP	Language Action Perspective
LCI	Lean Construction Institute
LPS	Last Planner System™
M	Matching
MDM	Multiple Domain Matrix
Mech.	Mechanical
NPD	New Product Development
NWD	Navisworks Document; Autodesk Navisworks is used to open a NWD file
OSHPD	Office of Statewide Health and Planning Development
PA	Project Alliancing
PANKOW	Pankow Builders

PD	Product Development
PDCA	Plan-Do-Check-Act
PE	Pacific Erectors
Plum.	Plumbing
PM	Project Manager
Q	Question
QA / QC	Quality assurance / quality control
REI	Rosendin Electric
Rep.	Representative
RLH	RLH Fire Protection
RVT	Revit; a file which can only be opened in Autodesk Revit
SG	Smith Group
SH	Sutter Health
SI	Southland Industries
SL	Silverman & Light
SNA	Social Network Analysis
SPS	Strategic Project Solutions, Inc.
TJ	Ted Jacobs Engineering
TVD	Target Value Design
VDC	Virtual and Design and Construction
VM	Van Mulder Sheet Metal, Inc.
VSM	Value Stream Map
VNGC	Van Ness and Geary Campus
WBS	Work Breakdown Structure
WPCS	WPCS International, Inc.

1 Introduction[1]

1.1 Background

The level of complexity is a critical dimension in characterizing projects, and construction projects are described as quick, uncertain, and complex (Howell and Ballard 1997). The general public tends to learn about the level of complexity of large-scale construction projects through the problems such projects run into, namely cost and schedule overruns and quality issues. These problems arise on well-known megaprojects[2] and on smaller projects alike. They are often related to the adaptation of project management practices to the characteristics of the specific project. Thus, project management includes the management of project-based complexity that originates in the project itself and in the project environment.

This dissertation focuses on the detailed design phase of construction projects, and detailed design itself is a complex process. During the detailed design phase of a construction project, the design organization develops concepts[3] that describe the final product of the project, the building. These concepts may be physical or abstract, and increasingly designers produce them using software tools and three-dimensional (or more) modeling. The design organization is a group of people, here called designers, who jointly carry out a series of tasks to design the building. This series of tasks makes up the design process, in which designers generate knowledge about the building. The design process is

---

1 Parts of this section have been published in Hickethier et al. (2012, 2013).

2 Lately, a number of German "megaprojects" appeared in the news because of cost and time overruns, for example the new Berlin Airport (BER), the Elbphilharmonics in Hamburg, and the Railway Project Stuttgart 21 (e.g., Schöttle and Gehbauer 2013).

3 Or "recipes" (Reinertsen 1997).

subject to uncertainty, because designers must decide on the characteristics of the final building as the process unfolds and infinitely many possibilities may exist. Design processes can be compared to problem solving: designers solve "wicked problems" – those that have "no definite formulation" and where there is no guarantee to find a solution (Rittel and Webber 1973). In order to complete a design task, the designer must generate knowledge (Hatchuel and Weil 2003). Uncertainty in design often surfaces through the need for iteration in the design process, during which the building design is reworked, refined, or improved (Wynn et al. 2007).

Designing a building requires a number of different skills and knowledge, typically provided by designers who work for different companies. The skills and knowledge needed may differ from project to project. Accordingly, designers on project teams tend to not have worked together before. Nevertheless, they must collaborate to generate design alternatives and decide on criteria to assess them, so as to achieve a design that delivers value to project stakeholders. In the process of learning about criteria and alternatives, designers and other project participants, e.g., design managers, exchange information, i.e., information flows between them.

Project participants must communicate with one another in order to exchange information. Communication can exist in several forms: verbally or graphically, digitally or paper-based, through plans, lists, or sketches, among others. The design process prescribes the flow of information between design tasks, and designers coordinate tasks through communication (Flores 1981; Macomber and Howell 2003; Maier et al. 2008; Pall 2000). Also, communication is a prerequisite for improvement as it is needed to provide information regarding results for reflection and analysis (Baecker 2003, p.21, 2006). The quality of communication among designers impacts task performance, and thus design process performance (Allen 1977; Chinowsky et al. 2008; Eckert et al. 2001; Tribelsky and Sacks 2010).

This dissertation addresses the problem of describing, analyzing, and improving communication in the detailed design phase of complex building projects. Detailed design builds on the schematic design of a facility and entails coordination and detailing of technical systems, while striving to improve customer value, e.g., through improved performance or cost reductions. Detailed design is also known as design development (American Institute of Architects 2007).

The commonly applied approach to the design of complex buildings is to divide-and-conquer: decompose the design problem into smaller parts (often called systems), and if these sub-problems are still too complex, further decompose them into even smaller parts (often called components). The purpose of decomposition is to split the problem into parts that are manageable by an individual or small group of people (Alexander 1964; Simon 1996). However, the resulting problem-parts are often interdependent, which causes the tasks designed to solve the problem-parts, as well as the generated solutions to be interdependent. Thus, the tasks and their solutions must be integrated, and choosing optimal solutions for the problem-parts does not necessarily lead to an optimal solution for the overall problem.

Detailed design can be seen as two interdependent sub-problems: (1) the sub-problem of 'what' to build (product design) and (2) the sub-problem of 'how' to build the 'what' (process design). In reality, these two sub-problems are assigned to two different groups of people: designers and builders, e.g., in the contract type Design-Bid-Build (DBB), referred to as "traditional project delivery" (Cushman and Loulakis 2001, p.6). This type of project delivery allows for competitive bidding to determine the contractor for construction, but it increases the risk of a lack of production knowledge ("how to build") while designing. In simple projects designers can often develop 'constructable' designs, because they have sufficient knowledge about the building process. However, they may miss opportunities for improvement of building design and construction process (Gil et al. 2000). The concept of integrating design and construction knowledge has long been integral to construction projects.

Historically, integration was embedded in the concept of the "master builder", who had sufficient knowledge to fill the roles of architect and builder at the same time (Cushman and Loulakis 2001, p.6).

Practitioners have recognized the need for and benefits of project integration. Concurrent Engineering (CE) proposes concurrent development of the 'what' and the 'how' during building design (Anumba and Evbuomwan 1997; Love and Gunasekaran 1997). Regarding integration, Lean Construction (Howell 1999; Koskela 2000; Koskela and Alarcon 1997) highlights the need for collaboration and continuous improvement (Ballard 2000a, 2008; Tsao et al. 2004). Two general trends emerged in the Architecture-Engineering-Construction (AEC) industry regarding integration:

(1) Organizational integration through contractual agreements. For instance, Integrated Project Delivery (IPD), Project Alliancing (PA), Design-Build (DB), and Design Assist (DA) integrate project participants across building systems and the building lifecycle. IPD and PA include an alignment of financial interest between project participants in order to foster collaboration. Zimina et al. (2012) show that IPD type contracts are beneficial in terms of project cost and schedule.

(2) Process integration through Building Information Models (BIM). BIM stores and supplies information for several processes, e.g., designing, estimating, and construction process planning, in one integrated database. The purpose of integration is to promote a shared understanding between project participants, to improve sharing of information, and to foster collaborative behavior.

Figure 1 shows the complexity of integrated AEC projects based on the framework of the Lean Project Delivery System (LPDS)[4] (Ballard 2008) (see section 2.3.2.2). Boxes represent the five phases of the LPDS along three dimensions: (1) building lifecycle, (2) building systems, and (3)

[4] Section 2.3.2.2 describes the LPDS.

project processes. Project integration demands not only an integration across the lifecycle of the building, e.g., Design, Supply, and Assembly, but also across Building Systems, e.g., Mechanical, Electrical, and Plumbing, and across project processes, e.g., Requirements Management, Design Optimization, and Trade Coordination.

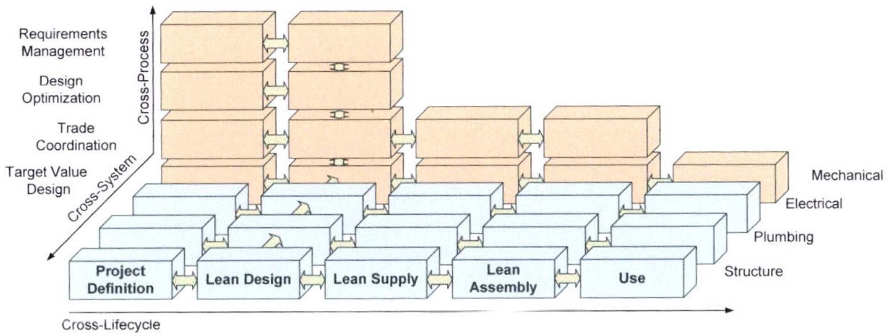

Figure 1: Cross-Lifecycle, Cross-System, and Cross-Process Integration in AEC Projects (based on Ballard (2008); Bergsjö et al. (2007))

1.2 Motivation

Project integration across systems and lifecycle increases the amount of knowledge and number of people participating in the design phase (as compared to DBB projects) (Thomsen et al. 2010a, p.11). But improved integration does not automatically reduce coordination deficiencies (Sherman 2004). Instead, project integration can increase coordination complexity, because a larger number of people must be coordinated. During the detailed design phase, coordination includes management of communication within the design team.

Figure 2 presents a preview of case study A (see chapter 4). The figure shows the communication structure between project participants of the Van Ness and Geary Campus (VNGC) hospital project[5], formerly known

[5] Section 4.1.2 describes the project.

as Cathedral Hill Hospital (CHH) project. This project applies an IPD-type contract, the Integrated Form of Agreement (Lichtig 2005) including Lean Construction methods. Project participants in the detailed design phase include both design and construction companies. Chapter 4 will discuss communication structures of the project in detail.

Different shades of blue represent the amount of information exchanged between companies (darker shades represent larger amounts of information flow). The information exchanges reveal a modular organization structure in which the Architecture Firm (SG) and the General Contractor (HB) serve as interfaces between the modules. Directly visible are the modules 'Owner' (CPMC/SH), 'Mechanical-Electrical-Plumbing' (SL/REI/TJ/SI), and 'Exterior' (KHSS/DE/PE/BAGS/DJ). The latter two modules consist of companies from two different disciplines: design and construction. The observed pattern of information exchanges shows integration across building systems and across the building lifecycle, but the pattern also reveals a high degree of complexity in the interaction between project participants.

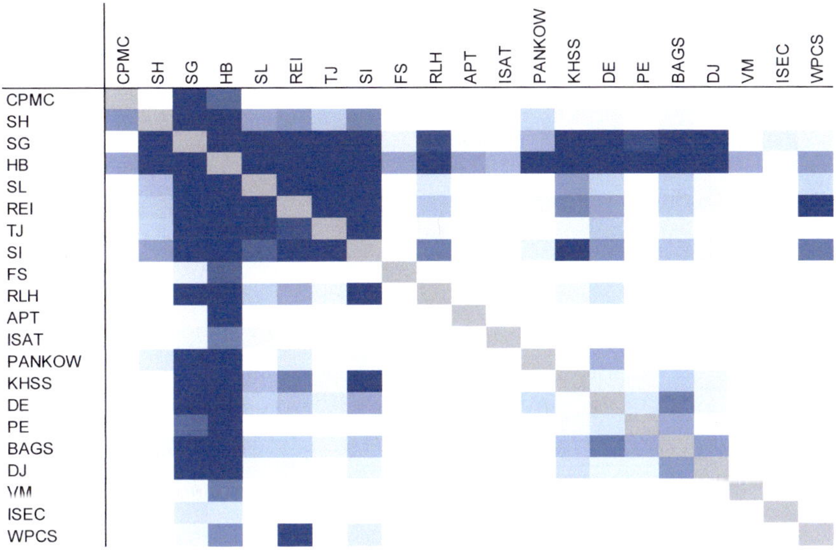

Figure 2: Information Exchanges between Companies at the Van Ness and Geary Campus (VNGC) Project in the Detailed Design Phase. Darker Shades represent larger Amounts of Information Exchanged. Read "Row Item receives Information from Column Item"

1.2.1 Intransparency of Information Flow in Design

Transparency of processes is a key to Lean Production Management, because it facilitates the implementation of mechanisms for control and improvement (Koskela 2000, p.63). The 'matter' of production is usually material, which is mostly visible. The 'matter' of design is often invisible and consists of information, and face-to-face communication or communication through media (or data carriers) transport information between designers. With the rise of digital communication and the use of 3D modeling, such as Building Information Modeling (BIM), communication between designers has become more invisible. Instead of sending a physical set of printed documents, designers now use integrated data-servers to access and alter building models. This development changes tracing of information through the production process of design. The flow of digital information is not as visible as the flow of physical document, thus means for achieving visibility of digital information flow are needed.

The structure of the design process impacts the structure of communication and vice versa. Iteration[6] of tasks is a characteristic of design processes, and iteration can be value-adding or wasteful (Ballard 2000b). Iteration in 'designing' may offer an opportunity for designers to deepen their understanding of the task and explore alternatives, so that they can deliver an outcome of greater value to the customer. This value-adding or so-called positive iteration is to be encouraged. Iteration is called wasteful, if it can be eliminated from the process without a loss of value or risking the success of the project. This so-called negative iteration (Ballard 2000) should be avoided.

Intransparency of communication between designers complicates the analysis of actual communication structures, which can serve as a starting point for analyzing iteration. This dissertation aims to develop a method for obtaining actual communication structures between designers and comparing it to planned communication. Chapter 5 will present a method for comparison of actual and planned communication.

1.2.2 Problems of Process Management in Design

Process modeling supports the management of information flow by achieving transparency of tasks and their dependencies; it supports project planning, it supports process coordination through execution and control, and it is the foundation of continuous improvement and learning about processes (Browning and Ramasesh 2007). Ineffective coordination of design processes often causes waste, either in the design phase or

[6] Iteration refers to the "repetition of nominally complete activities" (Ulrich and Eppinger 2004, p.16).

during construction. Scholars estimate that design and documentation problems cause between 45% and 70% of rework in construction (Jungwirth and Fuhr 1994; Love et al. 2008). During design, about 50% of iterations are wasteful (Ballard 2000b).

Deviating from planned processes causes problems, and root-causes for problems also originate in the design organization. According to Browning (2002), "the value of a process is compromised when information is "out of sync," forcing those who are executing activities to make assumptions in the absence of real information." Koskela (2000, p.198) states that "[...] in practice there are several factors tending to push the design process away from the optimal sequence", and that about half the disturbances originate in the design organization (Koskela 2000). Clarkson and Eckert (2005, p.70) explain that projects rarely compare models of planned processes to actually executed processes.

Short-cyclic tracking of commitments, e.g., with the Last Planner System (LPS) (Ballard 1994, 2000c), achieves transparency regarding fulfillment of planned process interactions. But it misses opportunities for process improvement, because it does not visualize the structure of information flow. Commitments for tasks can be kept, but include additional, unplanned iteration of information between these tasks. The LPS does not identify this structural misalignment between actual and planned information flow.

1.2.3 Opportunities for Achieving Transparency of Communication Structures

Researchers often achieve transparency of actual communication by using surveys to collect data, e.g., (Chinowsky et al. 2011; Kratzer et al. 2008; Morelli et al. 1995). This type of data gathering has also been applied for practical improvement purposes (Eppinger and Browning 2012, pp. 99ff.; McCord and Eppinger 1993; Sosa et al. 2004). Data collection through surveys is time-consuming and effort increases drastically as organizations grow. Therefore, it is not ideal for the implementation of quick cycles of continuous improvement through analysis of actual communication.

Increased use of digital communication also provides opportunities for achieving transparency of actual communication. In the field of business process modeling (BPM) Aalst (2005, 2011) measures process structures, i.e., what tasks were executed and how they were related. The approach "process mining" (Aalst 2005, 2011) discovers and collects data, which can describe actual interactions between people. Process mining maps an already executed process based on the traces it left – usually based on existing documentation of interaction, for example in logs of IT-systems. In the field of computational social sciences, Pentland (2012) collects data of actual interactions between people with "sociometric badges". People wear these badges and badges recognize the proximity to each other. Badges can, e.g., log durations during which they are within a certain distance to each other. These logs can then serve for modeling communication between people. Application of such technologies for obtaining indicators for actual communication facilitates data collection, thus reducing effort for building models of actual communication.

1.3 Research Objectives

The integrated organization of IPD projects provides opportunities for improving communication in the detailed design phase. Shifting away from the traditional silo-structure encourages people to structure their communication based on project needs. But as design organizations grow, some artificial boundaries between teams become necessary. For example, in IPD projects, membership in a team establishes one type of boundary. Organization design defines these boundaries. The structure of boundaries impacts how communication between people unfolds.

Misalignments between boundaries and actual communication requirements causes inefficiencies (Colfer and Baldwin 2010; Eppinger 2001).

Analysis of misalignments necessitates a model of actual communication. The first (1) objective of this dissertation is to show how models of actual communication can be obtained using project databases. Specifically, this research uses BIM to obtain data regarding actual communication between people. The second (2) objective of this dissertation is to show how obtained data can be used for improving communication structures. The third (3) objective of this dissertation is to analyze prerequisites for improving communication structures, and to check whether these prerequisites exist in IPD-type projects.

To approach objectives (1) and (2), this dissertation describes a method for improvement of communication structures using delta-analysis. 'Delta' refers to misalignments between structures of actual and planned communication structures. A set of metrics regarding misalignments facilitates analysis of communication. Transparency of the actual pattern of communication is prerequisite for the analysis of misalignments, and the integrated database of BIM offers opportunities for tracing communication digitally. Analysis of misalignments between actual and planned communication applies models based on the Design Structure Matrix (DSM)[7](Browning 2001; Steward 1981). The terms 'delta' and 'deviation' shall reflect an open perspective on misalignments between patterns of actual and planned communication. Neither pattern of communication is per definition the 'right' one. Instead, the purpose of delta-analysis is to find root-causes for deviations between patterns (see section 5.4.1.2 for a description of delta-analysis).

Based on the availability of the actual pattern of communication, there is need for a method which (1) enables evaluation of misalignments and (2) supports quick learning for process improvement based on the scientific method (see section 2.3.2.1). Quick learning loops can help

[7] Section 2.1.4.3 describes DSM.

avoid unwanted rework. Also, learning accelerates the integration of project participants' modeling processes into one holistic design process.

To approach the mentioned research objectives, this dissertation will:

- review and document current approaches for evaluation and improvement of communication structures,
- outline requirements for effective application of the method elaborated in this research and check whether these requirements are attainable on IPD type projects,
- develop a theoretical background and a procedural framework for application of the method, and
- document the application of the method on actual case study projects.

1.4 Research Questions

The research questions pertain to developing a method for improvement of communication structures using delta-analysis. The main research question is: how can (1) communication be made transparent in an efficient way such that (2) comparisons can be drawn between actual and planned communication in order to (3) continuously improve the design project?

Q1. How can a design team efficiently achieve transparency of actual and planned communication in the detailed design phase of a construction project?

Q2. How can the design team evaluate alignment of actual and planned communication? What are the metrics for evaluation?

Q3. How can the team use knowledge about misalignments between actual and planned communication to improve the design system continuously?

1.5 Research Approach

This dissertation describes two independent, but connected studies:

Study A: the goal is to examine, whether the prerequisites for application of the method exist in current AEC practice, specifically on projects applying IPD. Study A applies case study research to prove or discard a set of hypotheses regarding information flow at an IPD-project, the VNGC project. Evaluation of hypotheses employs metrics from Social Network Analysis (SNA), by which information flow is used to model the informal organization. To the knowledge of the author, no models of actual information flow between people in IPD-type projects exist. Thus, this case study contributes to the body of knowledge of IPD by checking whether proposed structural characteristics are actually in place. Case study research was chosen for this study, because the goal of this study is to examine the existence of requirements in current AEC design practice. A sample size of one case study is sufficient to show the attainability of prerequisites, if existence of these prerequisites can be shown.

Study B: The goal is to deliver a 'Proof of Concept' for the method itself. Study B applies a combination of case study research and constructive research. Two 'Proof of Concept' case studies were undertaken, (B1) the VNGC project and (B2) another large hospital project in California. Both projects apply BIM and both case studies investigate the modeling process of the interdisciplinary design team at the respective project. To the knowledge of the author, no case studies exist which use BIM data to model actual communication. Hence the case studies contribute a new kind of BIM application to the existing body of knowledge. Constructive research was chosen, because the goal of this study is to develop, improve, and test a method through 'Proof of Concept' experiments in AEC practice. A sample size of two case studies increases generalizability (Meredith 1998) and it allows cross-case analysis of patterns between case studies.

1.5.1 Case Study Research

Yin (2009) describes case study research as "an empirical inquiry that investigates a contemporary phenomenon in depth and within its real-life context, especially when the boundaries between phenomenon and context are not clearly evident." Eisenhardt (1989) describes case study research as "a research strategy which focuses on understanding the dynamics present within single settings." Case study results are often criticized, (1) because results may be influenced by the personal perception of the observer, i.e., the researcher, and (2) because results are not always reproducible (Meredith 1998). These two points of critique are connected: research results compromised by undocumented personal perception are hard to reproduce. A clear documentation of research objectives and boundaries as well as of the researcher's perceptions is necessary to achieve meaningful research results (Yin 2009, pp. 27ff.).

Benbasat et al. (1987) name three advantages of case study research: (1) it takes place in real life and delivers results for practice, (2) it focuses on understanding of phenomena by asking "why", and (3) it is appropriate for less mature fields of research where few prior studies exist. IPD and BIM are such fields of research.

1.5.2 Constructive Research

Constructive research, also called design science research, produces knowledge "through creation and implementation of a solution that is able to manipulate or alter a particular phenomenon" (da Rocha et al. 2012). Lukka (2003) describes constructive research as "a research procedure for producing innovative constructions, intended to solve problems faced in the real world and, by that means, to make contributions to the theory of the discipline in which it is applied." The construction in this research is the method for improvement of communication structures using delta-analysis, and the construction is tested and refined in two case studies. Development of the construction is iterative, and lessons learned in one case study are used to improve the construction for application in the following case study. Thus, constructive research often relies on case study research.

Constructive research can deliver practical and theoretical contributions (Lukka 2003). Evaluation of the practical contribution of the final construction regarding utility, quality, and efficacy (Hevner et al. 2004) is described in section 7.1 based on research questions. Contributions to theory are summarized in section 7.2 "Contributions to Knowledge".

1.6 What this Dissertation is not About

Process management and organization design literature both encompass large fields of knowledge, so it is important to delineate what is not part of this dissertation. The following issues are deliberately excluded from this dissertation:

- Execution of processes leads to results and structures of actual communication are only one type of result. The evaluation of processes pertains to process structures. Process evaluation does not include non-structural process results, e.g., time and cost performance.
- This dissertation focuses on integration from the perspective of interaction. Analysis of commercial terms regarding their effect on integration of project participants is not within the scope of this research.

1.7 Dissertation Structure

Figure 3 illustrates the dissertation structure. Chapter 2 reviews relevant literature. It focuses on fundamentals of complexity in the detailed design phase of AEC projects reviewing theories regarding systems and structures, design processes, and organization architecture. Also, chapter 2 reviews literature regarding production systems and lean construction.

Chapter 3 describes the research gap based on the fundamentals reviewed in chapter 2 and includes requirements for the method for improvement of communication structures using delta-analysis. Chapter 4 presents case study A, which tests through case study research whether the requirements from chapter 3 are attainable on IPD projects. Following, chapter 5 presents the method for improvement of communication structures using delta-analysis with its meta-model, an approach for data gathering, metrics for delta-analysis, and a procedural framework.

Chapter 6 presents case studies B1 and B2, in which the method for communication improvement using delta-analysis is applied. Case study B1 was completed at the VNGC project in San Francisco and case study B2 at a second project, both located in California, USA. The chapter closes with a cross-case analysis.

Chapter 7 summarizes the research findings, outlines the contributions to knowledge, and gives recommendations for future work.

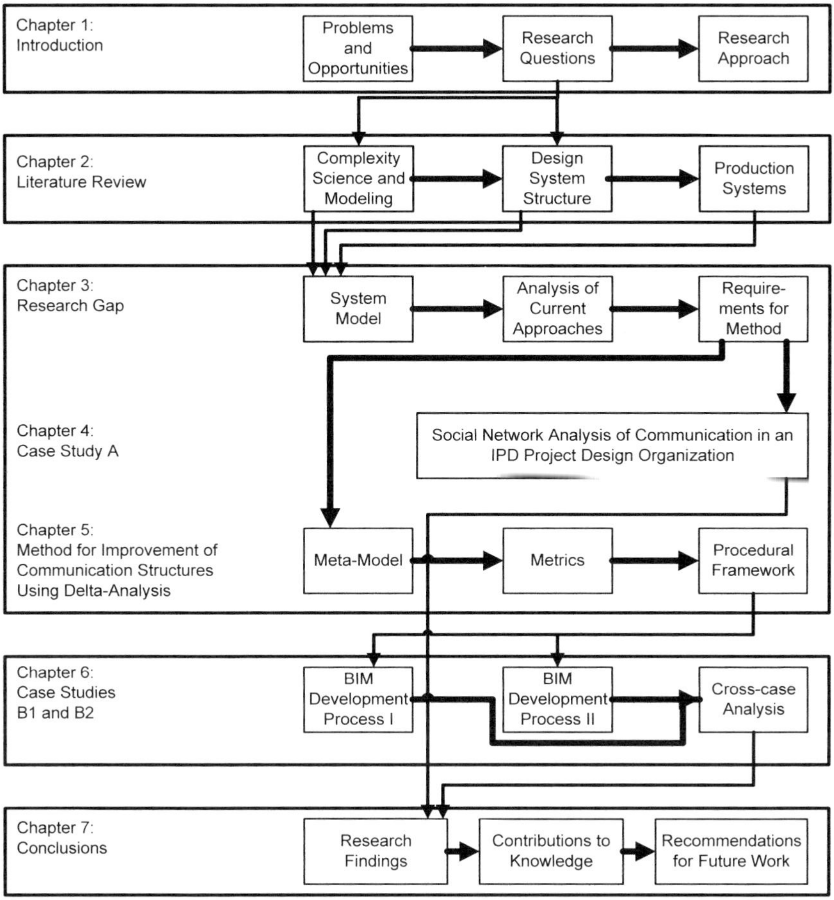

Figure 3: Dissertation Structure

2 Literature Review

The purpose of this literature review is (1) to describe the current state of the art in industry and academia in order to understand the foundation of the contributions to knowledge of this dissertation, and (2) to provide definition and vocabulary for the contribution.

Foundations from several scientific fields are the starting point for development of the method for improvement of communication structures using delta-analysis. These scientific fields were identified with the Design Research Methodology (Blessing and Chakrabarti 2009, pp. 63-65). This chapter explains each field in detail.

First, section 2.1 explains the origins of systems sciences, complexity and the related structures. This includes different means for modeling complex systems by formal description of structures. Based on these foundations, section 2.2 reviews the structural content of products, processes, and organizations in AEC design. Section 2.3 focuses on production theory, because production is in this dissertation assumed as the purpose of the design system. Section 2.4 summarizes the literature review. Several of these fields overlap and the purpose of figure 4 is to structure the fields for the literature review of this dissertation. Further, figure 4 also shows the contribution this dissertation aims to achieve, which is located in the joint analysis of organization architecture and process structure including contributions to process modeling.

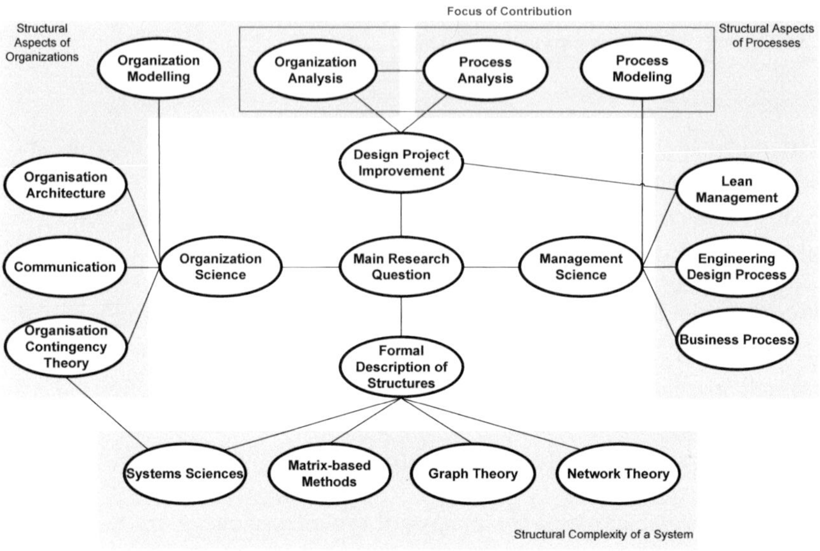

Figure 4: Relevant Scientific Fields to this Research

2.1 Complexity related to the Design Phase of Construction Projects

2.1.1 Types of Complexity

Complexity is widely regarded as one of the critical dimensions of projects (Baccarini 1996; Williams 1999). The term complexity is widely used in a large number of scientific fields, however, there is no agreement in the scientific community about a definition for the term (Horgan 1995). Also, there is neither an agreed upon definition for the term complexity in the scientific field of engineering (Piller and Waringer, 1999; pp. 5) nor in project management (Williams 1999).

Weaver (1948) first mentioned the term complexity in the field of cybernetics[8]. He defined "complexity" as the counterpart of "simplicity"; Kurtz and Snowden (2003) explain "simplicity" as the science of orderly systems, while complexity is the science of un-orderly and chaotic systems. Cybernetics (e.g., Weaver 1948), Systems Theory (Bertalanffy 1950), and Dynamic Systems Theory (Padulo and Arbib 1974) laid the foundation for Complexity Science[9]. The interest in complexity science surged in the 1970s and led to research in a large number of fields. The smallest common denominator between at least some of the fields is that the behavior of a system cannot be derived from knowledge about the characteristics of the constituent parts of the system[10]; instead the behavior of the whole is more than the sum of its parts. As Simon states: "given the properties of the parts and the laws of their interaction, it is not a trivial matter to infer the properties of the whole" (Simon, 1962, p. 468).

Simon's statement highlights two views on complexity: (1) the whole that consists of parts and connections between these parts and (2) the subject who defines what the whole is and who infers the properties of the whole. Schlindwein and Ison (2004) refer to these two views as (1) descriptive complexity and (2) perceived complexity. While descriptive complexity describes a characteristic of a system, perceived complexity represents the problems one encounters when trying to understand this system. Edmonds (1999, p.72) combines these two views in his definition of complexity:

"Complexity is that property of a model which makes it difficult to formulate its overall behavior in a given language, even when given

---

[8] Ashby defines cybernetics as the science of "Co-ordination, regulation, and control" of systems (Ashby 1956).

[9] Brian Castellani's map of complexity science at http://www.art-sciencefactory.com/complexity-map_feb09.html provides an overview of the development of complexity science over time.

[10] Section 2.1.1 provides a definition of "system".

reasonably complete information about its atomic components and their interrelations."

1. Descriptive Complexity

A multitude of ways exist to describe complex systems. Weaver (1948) introduced the distinction between (a) organized and (b) disorganized complexity. The distinction is rooted in the idea that the purpose of describing a system is to solve a problem. Rittel and Webber (1973) define a problem as the divergence between a current state and a desired state.

a) Problems of disorganized complexity can be described as "problem[s] in which the number of variables is very large, and one in which each of the many variables has a behavior which is individually erratic, or perhaps totally unknown" (Weaver 1948).
b) Problems of organized complexity can be described as "problems which involve dealing simultaneously with a sizable number of factors which are interrelated into an organic whole" (Weaver 1948).

Zamenopoulos and Alexiou (2005) propose three categories for descriptive complexity: (a) structural, (b) functional, and (c) behavioral complexity.

a) Structural complexity describes systems as consisting of parts that are structured in some way. For example, the power-law distribution describes that few parts are highly connected, while many parts are little connected to other parts.
b) Functional complexity expresses the difficulty in describing input to output relations of a system. For example, computational complexity theory describes the time and resources needed to complete a specific computation.

c) Behavioral complexity applies dynamic models to analyze systems. Systems are considered complex, when they show specific behaviors, such as emergence and self-organization. For example, multi-agent systems are used to model complex behavior of a system, which arises from interaction between agents, who operate on a few simple rules.

2. Perceived Complexity

Ashby (1973) highlights the importance of considering the observer's perspective on an object when discussing its complexity: "to the neuro-physiologist the brain, as a feltwork of fibers and a soup of enzymes, is certainly complex; and equally the transmission of a detailed description of it would require much time. To a butcher the brain is simple, for he has to distinguish it from only thirty other 'meats', [..].". Klir (1985) adds that "complexity is given a somewhat subjective connotation since it is related to the ability to understand or cope with the thing under consideration."

The relationship between the observer and the observed object constitutes perceived complexity. Thus, the level of perceived complexity depends on object, observer, and the characteristics of their relationship and the level of complexity is influenced by several factors: the attributes of the object and the subset of attributes, which the observer is interested in (a decision which is influenced by his/her goals), the knowledge and experience of the observer regarding these attributes, the resources and technique which the observer employs to increase his/her understanding of the attributes, and the characteristics of the object, for example whether the object is observable or dynamically changing over time.

Maturana and Varela (1987, pp. 21f.) describe that the observer's prior experience impacts his/her observation of the world, as they state "[..] we cannot separate our history of actions—biological and social—from how this world appears to us" (Maturana and Varela 1987, p. 23).

Hence, their (Maturana and Varela 1987) observations are based in the idea of constructivism, which states that what an observer perceives as reality is only a construction in the observers mind.

Edmonds (1999) explicitly refers in his definition of complexity to the complexity of a model. Thus, his definition of complexity includes the fact that perceived complexity is not the complexity of the real world, but instead it is the complexity of the subjective model that the observer develops based on his observation of the world.

Kurtz and Snowden (2003) highlight the need to align people's views on the nature of the problem at hand. They argue that different classes of problems demand different strategies for solving the problem. The first step of problem solving then becomes to achieve a common understanding on the quality of the problem among group members.

Rittel and Webber (1973) distinguish between 'tame' and 'wicked' problems. Tame problems can be solved with a linear process; Weaver (1948) describes this class of problems as "problems of simplicity". Wicked problems do not have a "definitive formulation", each problem is unique, it is not possible to formulate a problem description unless a solution is available, and there is no 'right' solution to a wicked problems, but rather good or bad ones. Conklin and Weil (1997) describe the challenges of solving wicked problems in groups of people. They recommend integration of all stakeholders of the problem and an iterative learning process, which consists of two main steps: (1) analysis of the process and (2) synthesis of a solution. All group members shall work on the same step at a given time. Further, they recommend structured documentation of problem requirements, criteria for evaluation of solutions, and development and documentation of possible solutions throughout the process. This stringent documentation helps to develop a shared understanding in the group of people, who work on solving the problem.

2.1.2 Definitions and Characteristics of Structural Complexity

Understanding complex systems demands a terminology, which is provided by the definitions of model, system, entities and relations, structure, domain, uncertainty, and ambiguity. These definitions draw from and build upon the approach "Structural Complexity Management", first described by Maurer and Lindemann (2007) and Maurer (2007), and further refined by Kreimeyer (2009) for the engineering design processes.

- Model

Modeling serves the purpose of analyzing and better understanding a system (Browning 2002). Stachowiak (1973, pp. 131f.) names three properties for a model, (1) mapping property, (2) reduction property, and (3) pragmatic property.

(1) Mapping property: a model is a representation of a real or fictitious original entity. Both model and original entity have attributes, and modeling maps attributes of the original entity to attributes of the model.

(2) Reduction property: models usually include only a subset of the attributes of the original entity. The subset consists of attributes that are relevant to the developers and/or users of the model.

(3) Pragmatic property: developers chose the subset of attributes with a goal in mind at a specific time. Thus, when applying an existing model, it is important to consider the original purpose of the model, the time when it was built, and who the model was built by and for. Models are a substitute for the original entity and these considerations limit the applicability of a model.

Stachowiak (1973, p.129) distinguishes between descriptive and prescriptive models of an original entity, where descriptive models represent a current state and prescriptive models represent a desired state of the original entity.

Mendling (2008) criticizes Stachowiak's perspective on modeling, because Stachowiak neglects that development of the model itself is "heavily influenced by the subjective perception of the modeler"; he further criticizes that Stachowiak's perspective is rooted in positivism instead of constructivism (Mendling 2008, p.7). Perceived complexity of reality influences the modeler when observing the original entity, thus only the modeler's perception is the basis for the model. Mendling (2008, p.8) argues that this characteristic of modeling demands quality criteria and he recommends the "Guidelines of Modeling" by Becker et al. (1995):

- System correctness: the model is syntactically and semantically correct,
- Relevance: only the parts of interest of the original entity are mapped to the model,
- Economic efficiency: the trade-off between the effort for developing the model and making it as complete as possible,
- Clarity: to ensure that a user is able to understand the model,
- Comparability: the consistent utilization of guidelines in a modeling project, e.g., naming conventions,
- Systematic design: the clear distinction of different views on the original entity.

Mendling (2008, p.8) recommends the definition of a modeling technique to attend to the guidelines in a modeling project.

- System

Kreimeyer (2009, p.40) defines a system as:

"a set of entities of (possibly) different types that are related to each other via various kinds of relations. The system is delimited by a system border, across which inputs and outputs of the system are possible as an interaction with the environment. The system fulfills a purpose, which guides the meaningful arrangement of entities and relations. The behavior of the system is, in turn, due to the arrangement of the system's elements."

This definition is based on prior work by Lindemann (2009, p.336) and Wasson (2006, p.18). Wasson (2006, p.18) specifically mentions that entities work "synergistically to perform value-added processing". Chu et al. (2003) describe that the definition of a system splits the world into a system and its ambiance. They highlight the importance of considering the context of the system, i.e., the interaction of the system with its environment. Kreimeyer's (2009) definition integrates Wasson's (2006) call for value adding through the demand that a system shall "fulfill a purpose", and it integrates Chu et al.'s (2003) call for considering the system's environment. Hence, this research adopts Kreimeyer's (2009) definition of a system.

Several properties of systems have been described in related literature. Table 1 presents an overview.

Table 1: System Properties (based on Baldwin and Clark (2000, p.63); Kirsch (2009, pp.13f.))

Technical systems	Natural systems
Developed by humans, e.g., machines, buildings, software	Evolved through self-organization, e.g., living organisms, social groups
Development follows a plan	Development follows rules

Socio-technical systems Systems with technical and social elements, e.g., companies	
Complicated systems Many different but static elements Many different but static relations System behavior is constant and predictable	Complex systems Elements can change their properties Relations can change their properties System behavior is variable and unpredictable
Static systems System state does not change over time	Dynamic systems System state changes over time
Closed systems No relations with other systems / the environment exist	Open system Relations with other systems / the environment exist
Purpose-oriented systems The system serves a certain function in alignment with the interests of the system's environment The purpose can be deduced only by observing the system from the outside	Goal-oriented systems The system defines its own goals The system strives to attain these goals by itself
Deterministic systems System behavior is completely predictable	Probabilistic systems System behavior is not completely predictable (Table continued on next page)
Modular system In the structure of the system more than one group of elements exists in which elements are highly related. Relations between groups of elements are sparse	Integrated system No group of highly connected elements exists, which is sparsely connected to the rest of the system

- Entities and Relations

System structures consist of entities and relations. Several fields of research apply similar concepts, for instance Graph Theory and matrix-based methods such as DSM and MDM. Kreimeyer (2009, p.41) provides an overview of the terminology in different fields. Table 2 presents terminologies for entities and relations in Systems Theory, Graph Theory, Network Theory, and Design Structure Matrix / Multi Domain Matrix literature.

Table 2: Terminology to describe Parts of a System (expanded based on Kreimeyer (2009, p.41))

Term in Systems Theory	Entity	Relation
Term in Graph Theory	Vertex	Edge, arc
Term in Network Theory	Node	Link
Term in Design Structure Matrix / Multi Domain Matrix literature	Element	Relation, dependency (often implies direction)

- Structure

Maurer (2007, p.32) describes a system's structure as "the network formed by dependencies between system elements and [it] represents a basic attribute of each system. Structures can be characterized by the specific compilation of implied linkages between system elements and can be divided into subsets."

One important structural property of a system is modularity. Modularity regards the group structure of a system where a group consists of one or more elements. A system is considered modular, if more than one group of elements exists in which elements are highly related and relations between groups of elements are sparse (Baldwin and Clark 2000, p.63). These groups are called modules.

- Domain

Systems can contain several types of entities, e.g., people or documents, which are connected by relations of different natures, e.g., commitment (to person) or citation of (document). Within one domain, entities as well as relations have similar meanings. Thus, domains sort entities and relations into "homogenous networks" (Maurer 2007, pp.71f.), which enables efficient and purposeful analysis of large systems (Kreimeyer 2009, p.41).

- Uncertainty

Tushman and Nadler (1978) define 'uncertainty' as "the difference between information possessed and information required to complete a task." Schrader et al. (1993) further specify uncertainty from a structural perspective in their distinction between uncertainty and ambiguity. Ambiguity is a lack of clarity regarding the structure of a system: information about relations between entities of a system is missing or not all entities of a system are known, which in turn causes a lack of information about their relations. Uncertainty is a lack of information regarding the attributes of the entities of system, when the structure of a system is known. According to this definition, ambiguity can cause uncertainty. Ambiguity and uncertainty are attributes of a system's structure. Pich et al. (2002) add perceived complexity to Schrader et al.'s (1993) distinction between ambiguity and uncertainty. Here, perceived complexity is high when a great number of entities are intensely related.

Uncertainty can also be specified by its source:

- product-related uncertainty: Albers and Meboldt (2007) describe product-related uncertainty which pertains to ends and means of the product under development. The purpose of the design process is to reduce product-related uncertainty by generating knowledge regarding ends and means. A reduction of product–related uncertainty causes an increase in product specificity.

- process-related uncertainty: Russell (2013, p.2) describes process-related uncertainty which pertains to a lack of assurance or reliability of process results, i.e., the gap between what was planned and what actually happened. A reduction of process-related uncertainty causes an increase in process predictability.

- external uncertainty: Open systems interact with their environment, thus external uncertainty resulting from, e.g., changes in code requirements or legislation, can impact the project.

It should be noted that these three categories of uncertainty can impact each other.

2.1.3 Complexity in the AEC Industry

Several scholars have researched sources and characteristics of complexity in projects and specifically in construction projects.

2.1.3.1 Project Complexity

The Project Management Institute (PMI) defines a project as "a temporary endeavor undertaken to create a unique product, service, or result" (Project Management Institute 2008, p.442). Baccarini (1996) characterizes project complexity by the number of parts and the variety of parts in a system; thus complexity is not directly equal to size. He identifies organization, technology, environment, information, decision making, and systems as sources of complexity. Williams (1999) extends the characteristics listed by Baccarini (1996) with uncertainty, which encompasses stochastic effects and missing information. Further, Williams (1999) adds uncertainty in goals to sources of complexity, and states that uncertainty can spread across sources of complexity. Vidal and

Marle (2008) characterize project complexity by size, variety of parts, interdependence of parts, and the context of the project. Further, they identify organization and technology as sources of complexity and they develop a framework, which describes detailed factors that constitute these sources. Remington et al. (2009) provide a framework for project complexity in which they name difficulty, non-linearity, uniqueness, communication, context dependence, clarity, trust, and capability as characteristics of complexity. They name goals, means to achieve goals, number and interdependency of parts, timescale of project, and environment (market, political, regulatory) as sources of complexity. Geraldi et al. (2011) provide a broad literature review and develop yet another framework for project complexity. They provide characteristics of complexity: socio-political complexity, pace, dynamic, uncertainty, and structural complexity. Further, they present sources of complexity for each characteristic (table 3).

Table 3: Characteristics and Sources of Project Complexity (Geraldi et al. (2011))

Characteristic	Source
Socio-political complexity	Importance of project; support to project from stakeholders; fit/convergence of opinions, interests and requirements; transparency of hidden agendas
Pace	Pace
Dynamic	Change
Uncertainty	Novelty, experience, availability of information
Structural complexity	Size, variety, interdependence

2.1.3.2 AEC Project Complexity

According to Baccarini (1996), construction might be the most complex process in any industry. Bertelsen (2003) sees evidence for complexity in plan failure, delays, cost overruns and grief. Howell et al. (1993)

identify uncertainty in project goals as a source of complexity. Gidado (1996) names several sources of complexity in the AEC industry: demand for speed in construction, cost and quality control, safety in the work place and avoidance of disputes, technological advances, economic liberalization and globalization, environmental issues and fragmentation of the construction industry. Dubois and Gadde (2002) also name several sources for complexity in the AEC industry: number of technologies and interdependencies between them, rigidity of sequence between the main operations, overlap of process elements or stages, lack of complete activity specification, unfamiliarity with local resources and local environment, lack of uniformity of materials, work, teams with regard to time and place, and unpredictability of environment. Bertelsen (2003) names conflict of interest between the project owner and project participants as a source of complexity.

These different frameworks and sources highlight the extent of project complexity and the variety of its sources identified in prior literature. It is not the goal of this research to provide another framework for classification of sources of complexity, but instead to underline the need for management of complexity.

2.1.4 Formal Description of Structures

Several methodologies for the modeling of complex structures exist. This section provides an overview over existing methodologies. This overview does not aim at being complete, but rather to describe the methodologies that are of interest to this research.

2.1.4.1 Systems Theory

General systems theory (GST) (Bertalanffy 1950) is seen as the origin of systems science. Systems science deals with the behavior of systems. It describes relations between entities with differential equations and it assumes systems as open, meaning that they interact with their environment (Bertalanffy 1950). The purpose of GST is to provide an overarching theory of systems across different fields of science. GST proposes four principles (Probst 1987, p.76):

(1) Complexity: the structure consists of entities and their relations, which can both change dynamically.
(2) Self-reference: behavior of the system affects the system itself, thus possibly changing system behavior.
(3) Redundancy: it is not possible to identify controlling entities, because they cannot be separated from the entities being controlled.
(4) Autonomy: system behavior is (only regarding a subset of attributes) independent from the system environment.

Pulm (2004, pp.22f.) describes two paradigm shifts in the history of systems theory:

(1) The first shift from systems theory to (first order) cybernetics introduced the concepts of open systems and self-organization. Self-organization is related to self-reference and it describes how a system structures itself from influences created in itself. Hence, a system can emerge through self-organization.
(2) The second shift from (first order) cybernetics to second order cybernetics applies the concept of constructivism to systems theory and it introduces the concept of autopoiesis. The application of the concept of constructivism resulted in integrating the observer and the system he/she observes: the observer becomes part of the system, because his view of the system impacts the way he/she understands and interprets it. Autopoiesis refers to the concept of a system being able to reproduce its own entities from its existing entities, i.e., the system exists in an environment and it survives and adapts within the environment (Maturana and Varela 1987, pp.43ff.).

Being part of second order cybernetics, Checkland (1989) introduces the soft systems methodology that acknowledges constructivism: when people with their subjective views interact to solve a given problem, soft systems methodology proposes learning about the problem properties. The term 'soft system' indicates a not well-defined problem, as compared to 'hard systems', in which the problem is well-defined. Hence, the problem definition for a soft system is emerging over time, and collaborative, participatory debate can foster learning about the problem (Checkland 1989).

2.1.4.2 Graph Theory

Graph theory is a method for modeling and analyzing the relations between entities. Two finite sets, vertices (entities) and edges (relations), define a graph: $G = (V, E)$. Both, vertices and edges can have additional attributes, e.g., weightings (Gross and Yellen 2005, pp.1f.). Graph theory serves the analysis of a large number of different network types.

Graph theory is a generic modeling method for networks, and networks can have the following basic properties (Newman 2003, p.3):

- Networks can have one or more different types of vertices and one or more different kinds of edges.
- Edges can be directed ("digraph") or undirected.
- Edges can have a weight or be unweighted.
- Directed networks can be cyclic, i.e., containing closed loops of links.
- An edge can connect a node to itself ("loop").
- Vertices can have multiple links between them ("multigraph"), or one link connecting one node to many others ("hyperedge").

Apart from analysis of structures, graph theory can also visualize structures. Usually graphs are depicted as boxes (vertices) and arrows (edges). However, several arrangements of vertices are possible and the layout of a graph impacts reception by the observer. Hence, algorithms have been developed for arranging graphs.

Several approaches for graph visualization exist (Battista et al. 1998). Force directed graphs, also called Spring Layout (Fruchterman and Reingold 1991), have the advantage of providing an intuitive layout (Battista et al. 1998, p.29). The visualization algorithm models the graph layout as a system of entities with forces acting between them, and the algorithm aims at finding a layout with minimal energy in the system. Force-directed algorithms use information from the system itself to calculate the layout (Kobourov 2013, p.383). For instance, algorithms can aim at laying out the Euclidian distance between a pair of vertices proportional to the number of vertices on the shortest path between these two nodes (Battista et al. 1998, p.312). Recent work applies several centrality measures (see section 2.1.4.4) to approximate the Euclidian distance between nodes (Bannister et al. 2013). Many force-directed algorithms deliver similar visualizations (Battista et al. 1998, p.324), however force-directed algorithms only deliver useful results for graphs with less than a few hundred vertices (Kobourov 2013, p.384).

Battista et al. (1998) provide an example of the arrangement of a force-directed graph. Figure 5 "shows a graph where vertices have been replaced with electrically charged particles that repel each other and edges have been replaced with springs that connect the particles. An equilibrium configuration, where the sum of the forces on each particle is zero, is illustrated in [graph b) of figure 5]. This configuration can be interpreted as a straight-line drawing of the graph, as in [graph c) of figure 5]" (Battista et al. 1998, p.303).

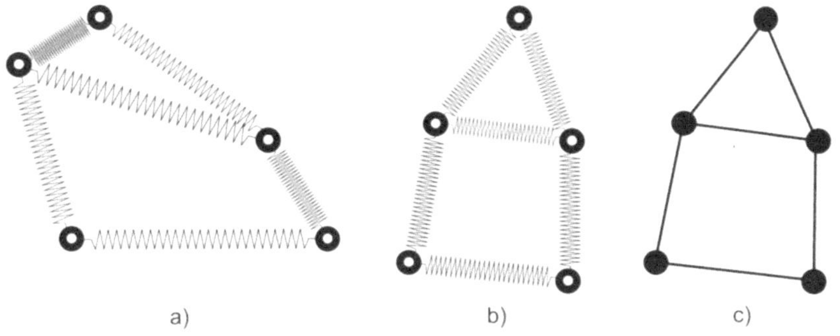

Figure 5: Formation of a Force-directed Graph (Lindemann et al. 2008) based on Battista et al. (1998)

2.1.4.3 Matrix-based Methods

In his essay "The Architecture of Complexity" Simon (1962) analyzes complex structures also applying square matrices for denoting the influence elements have on each other. Steward (1962) applies square matrices to analyze structures of equations. Vester (2002, p.165) applies a square matrix called "Papiercomputer" to analyze cause and effect relationships between elements of a system. In the field of Systems Engineering, which is rooted in Systems Sciences, Lano (1977) develops the N^2-Matrix to model interfaces between elements of a system. Steward develops the Design Structure Matrix (DSM) (Steward 1981) method to better plan projects that involve many interdependent variables. Eppinger and Salminen (2001) propose inter-domain analysis of several DSMs. Yassine et al. (2003) introduce connectivity maps that connect DSMs by establishing relations between elements from the different domains as represented in each DSM. Danilovic and Browning (2004) add Domain Mapping Matrices (DMM) to the DSM modeling method. Maurer (2007) further develops the modeling method by introducing Multi Domain Matrices (MDM).

Browning (2001) distinguishes four types of DSM: (1) component-based and (2) team based DSM, which are both static DSMs, and (3) task-based and (4) parameter-based DSM, which are dynamic (figure 6). Static DSMs capture the state of a system at a specific point in time, i.e., all elements and relations exist simultaneously. Dynamic DSMs capture elements and relations of a system, which are created and terminated over time, thus not all exist at the same time.

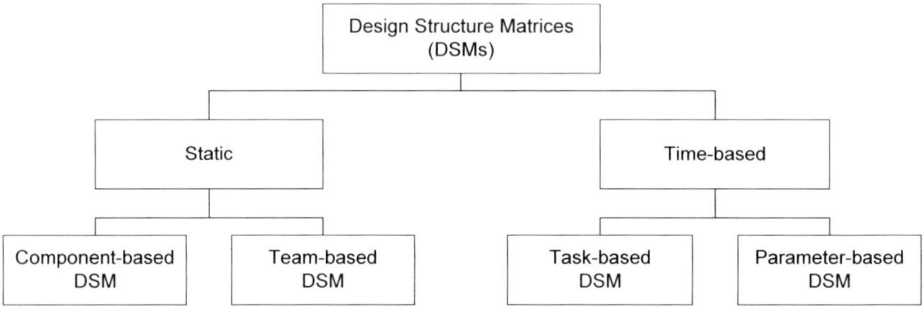

Figure 6: DSM Taxonomy (Browning 2001)

Figure 7 shows a simple process and the related binary DSM. Element names, in this example 'tasks', are shown across the top and the left side of the matrix in equal order from left to right and from top to bottom. In the center of the matrix, markings denote relations between elements, in this example the mark "X" stands for an output-input relation between two elements: for instance, the output of task one is the input for task two. Several types of marks have been used in binary DSMs, e.g., "X", "1", or "•". Also, DSMs can portray numeric dependencies instead of binary dependencies (e.g., (Browning and Eppinger 2002; Pimmler and Eppinger 1994)).

Elements of a DSM can by definition not have reflexive relations. Hence the diagonal of the matrix always stays empty. Two different notations for DSM exist: upper and lower diagonal. Upper-diagonal DSMs follow a row-to-column logic - if the row element precedes the column element, the field of the matrix on the intersection between row and column contains a mark. Lower-diagonal DSMs follow a column-to-row logic - if the column element precedes the row element, the field of the matrix on the intersection between row and column contains a mark. The names upper and lower diagonal DSM stem from the fact that sequenced matrices[11], which contain only feed-forward relations will only show marks either above ("upper") or below ("lower") the diagonal of the matrix. Both logics can be transferred into each other by transposing the matrix. The example in figure 7 shows an upper-diagonal DSM, and this logic is also applied throughout this dissertation. The arrow in the upper left hand box signals that the upper-diagonal definition is applied. The example in figure 7 refers to a simple, iterative design process of a house that includes a foundation, walls, and a roof. A specific task completes the design for each of these three parts. This design process is assumed to be iterative, because design of the house will begin with first drafts of each of the parts and then be followed by iterations to refine each part within the constraints of the overall house design.

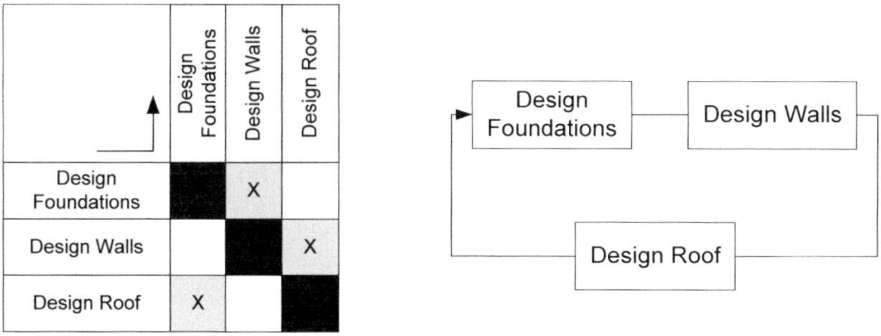

Figure 7: Binary DSM of a Simple Process

Domain Mapping Matrices (DMMs) extend DSM modeling by representing relations between elements of different domains. Figure 8 provides an example of task responsibilities by people: relations between tasks and people are shown in the matrix. In this example a mark in the matrix

[11] Sequencing is presented in the following section.

represents a person's responsibility for completing a task from figure 7. DMMs can be binary or numerical (Kreimeyer 2009, p.49).

	Design Foundations	Design Walls	Design Roof
Foundation Engineer	X		
Wall Engineer		X	
Roof Engineer			X

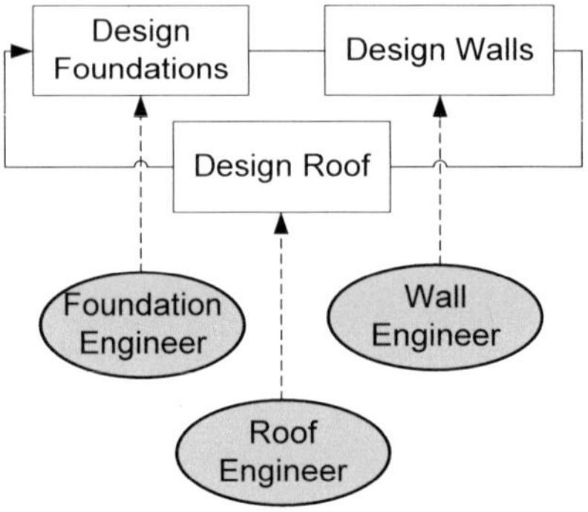

Figure 8: Binary DMM for the Process in Figure 7

Multi-Domain Matrices (MDMs) combine DSM and DMM into a framework. Maurer (2007, pp.57f.) structures and generalizes existing DSM and DMM methods by integrating super-diagonal and sub-diagonal DMMs with DSMs. Hence, MDMs can show directional as well reciprocal relations between elements from different domains. Further, each domain can consist of one or more matrices. The MDM approach enables modeling of systems that include different types of elements and relations by grouping them into domains and modeling dependencies between elements of different domains. Figure 9 shows an example based on previous figures 7 and 8: the super-diagonal DMM is empty and one additional DSM for the people domain is introduced.

	Design Foundations	Design Walls	Design Roof	Foundation Engineer	Wall Engineer	Roof Engineer
Design Foundations		X				
Design Walls			X			
Design Roof	X					
Foundation Engineer	X				X	
Wall Engineer		X				X
Roof Engineer			X	X		

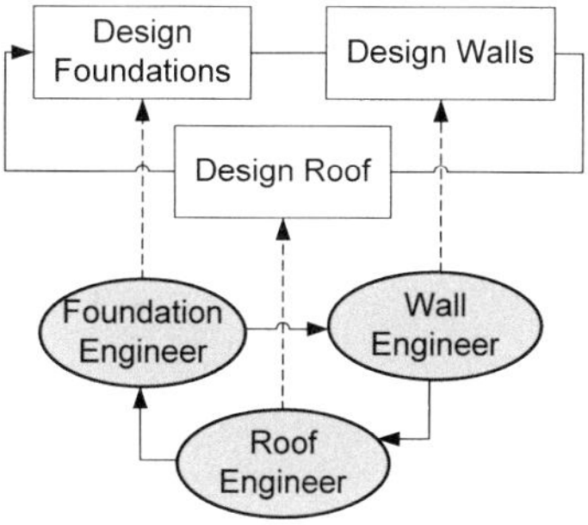

Figure 9: Multi Domain Matrix combining DSM from Figure 7 and DMM from Figure 8 and introducing one additional DSM in the People Domain

Maurer (2007, p.82) extends the MDM modeling approach by providing logics for computing DSMs by mapping relations across domains. Maurer identifies six cases for computing aggregate DSMs from existing native information in the form of DMMs and DSMs. Figure 10 shows the six cases based on the example of people working on documents. The goal of all six cases is to compute the people DSM: the relations which are aggregated from existing native information are shown as dashed connections between people-icons.

Case 1 uses the super-diagonal people-documents DMM: people who work on the same document are connected to each other in the people domain. The computed relation is reciprocal, because only information on accessing the document is provided and a direction of dependency cannot be inferred from this information. Multiplication of the super-diagonal DMM with the transposed DMM computes the people DSM.

Case 2 uses the sub-diagonal people-documents DMM: people who require the same document are connected to each other in the people domain. The computed relation is reciprocal. Multiplication of the sub-diagonal DMM with its transposed self computes the people DSM.

Case 3 uses both super and sub diagonal DMMs. Joining information regarding (1) what documents people work on and (2) what documents people require for their work enables computation of directed dependencies, which is indicated by the dashed arrow in figure 10. Multiplication of the super-diagonal DMM and the sub-diagonal DMM computes the people DSM.

Case 4 uses the documents DSM and the super diagonal people-documents DMM. Person A and person B work on different documents 1 and 2, and these documents are related: document 1 is an input for document 2. Directed dependencies in the documents domain enable computation of directed dependencies in the people domain. Multiplication of the super-diagonal DMM with the documents DSM and the transposed super-diagonal DMM computes the people DSM.

Case 5 applies a similar logic as case 4. Here, the sub-diagonal DMM is applied to compute directed dependencies between people instead of the super-diagonal DMM. Multiplication of the sub-diagonal DMM with the documents DSM and the transposed sub-diagonal DMM computes the people DSM.

Case 6 uses the maximum of native matrices by aggregating super-diagonal DMM, sub-diagonal DMM, and documents DSM to compute the people DSM. Multiplication of the super-diagonal DMM, sub-diagonal DMM and the documents DSM computes the aggregate people DSM.

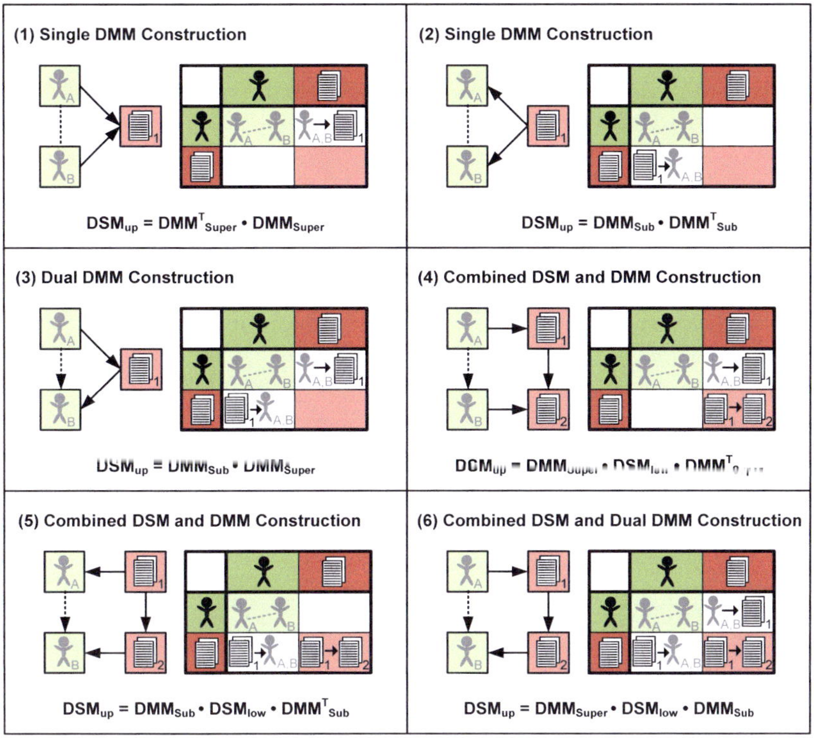

Figure 10: Computation of DSM from MDM subsets (based on Maurer (2007, pp.82ff.))

Kreimeyer (2009, p.51) structures techniques for DSM analysis as follows: "there are several strategies to analyze the DSMs generated. Classically, a DSM is used for sequencing, tearing, banding, and clustering. In sequencing, the rows and columns of a flow oriented DSM are rearranged in a way that as few relations as possible remain below the diagonal, thus reducing the number of active feedbacks, leading to an ideal sequence. However, such an ideal sequence cannot always be found. Tearing consists of choosing the set of feedback marks that obstruct sequencing the DSM. The relations that need to be removed are called 'tears'. Banding rearranges the rows and columns in a way that blocks of parallel entities remain, which, for example, in a process can be executed independently of each other. Thus, a 'band' represents a group of elements being active in parallel. Clustering is executed to find those clusters of entities that are mutually related."

Figure 11 shows the concept of each of the four classic techniques. Maurer provides detailed descriptions of the techniques (Maurer 2007, pp. 225-239).

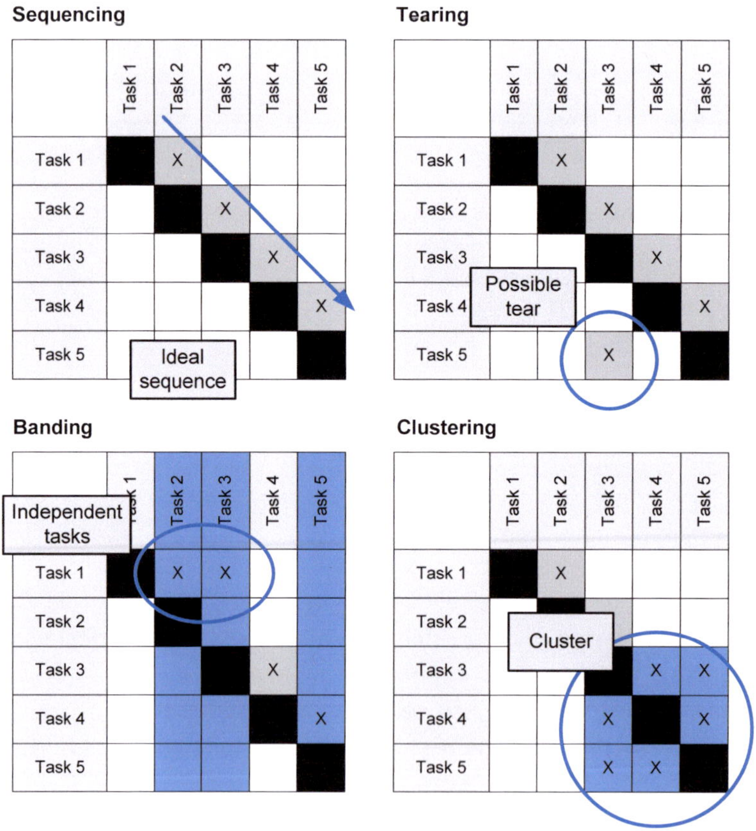

Figure 11: Classic DSM Analysis Techniques (Kreimeyer 2009, p.51)

Additionally to the classic analysis techniques, Maurer (2007, pp.225-239) provides a number of structural characteristics for the analysis of MDMs. Kreimeyer (2009, p.52) defines a structural characteristic as "a particular constellation of entities and relations, i.e., it is a particular pattern formed from nodes and edges in the graph. The characteristic gains its meaning by the way the pattern is related to the actual system it is part of, i.e., it must serve a special purpose in the context of the overall system. A structural characteristic only possesses significance in the context of the system it is describing." Kreimeyer (2009, p.52) categorizes existing structural characteristics by number of nodes and edges and provides graphic examples.

Classic DSM analysis techniques and structural characteristics are useful to analyze one existing DSM. De Weck (2007) introduces the delta-DSM, which subtracts one DSM from another in order to yield the structural difference regarding relations between two DSMs (figure 12). Eben et al. (2008) extend the delta-DSM definition by also allowing introduction and elimination of elements in order to model system change over time.

DSM (n+1) = DSM (n) + Delta-DSM

DSM (n+1)

	Task 1	Task 2	Task 3
Task 1		X	
Task 2			X
Task 3	X		

DSM (n)

	Task 1	Task 2	Task 3
Task 1		X	X
Task 2			
Task 3	X		

Delta-DSM

	Task 1	Task 2	Task 3
Task 1			-X
Task 2			X
Task 3			

Figure 12: Delta-DSM

DSM methods have been applied in the AEC industry in several documented case studies on a range of projects. Tuholski (2008, p.70) provides an overview.

2.1.4.4 Network Theory

Network theory is similar to graph theory and builds on graph theory. Network theory describes cases from the real world while graph theory describes theoretical cases (Barabási 2015, p.26). While graph theory analyses the impact and position of specific vertices and edges, network theory applies statistical measures for the analysis of large graphs (Newman 2003, p.2). Network theory has three goals (Newman 2003, p.2):

(1) To search for statistical properties that describe the structure and behavior of a system through a measure.
(2) To create models of networks that augment the meaning of the statistical measures.
(3) To predict the behavior of networks based on statistical measures and rules regarding the behavior for specific vertices.

Several network properties exist. Table 4 provides an overview of important properties.

Table 4: Common Network Properties (Newman 2003, pp.10f.)

Size of network	number of nodes and number of edges
Mean degree	mean number of edges per node
Mean distance between two nodes	mean number of nodes one has to traverse to travel between a pair of nodes
Diameter / longest geodesic distance	longest of all shortest paths between a pair of nodes
Network density[12]	number of existing triangles divided by number of possible triangles in a complete graph

[12] Also known as 'Clustering Coefficient' but not to be confused with 'Clustering' of a DSM.

Networks can contain clusters of highly connected nodes, where the connectedness between clusters is low. If two clusters are not connected to each other at all, they are called 'components' of a network. Cluster structures, also known as community structures, can be identified using cluster analysis algorithms (Newman 2003, p.17). Several algorithms for community identification exist (Newman 2003, pp.18-19), and algorithms exist that can identify structures of overlapping communities (Palla et al. 2005).

Networks can be represented as adjacency matrices (Barabási 2015, p.39). The adjacency matrix can be seen as a type of binary DSM; in case of directed links the adjacency matrix is an upper diagonal DSM. Hence, clusters can also be identified by blocks of element groups along the diagonal of the matrix (figure 11). However, it must be noted that adjacency matrices allow reflexive relations while DSMs do not include reflexive relations.

A commonly used property of nodes is 'centrality'. Wasserman and Faust (1994, p.178) distinguish degree, betweenness, and closeness centrality. Figure 13 shows examples for each type of centrality.

(1) Degree centrality represents the number of nodes a node is directly connected to.
(2) Betweenness centrality represents the number of shortest paths between any two other nodes in the network that a specific node is part of.
(3) Closeness centrality represents the distance of a node to other highly connected nodes.

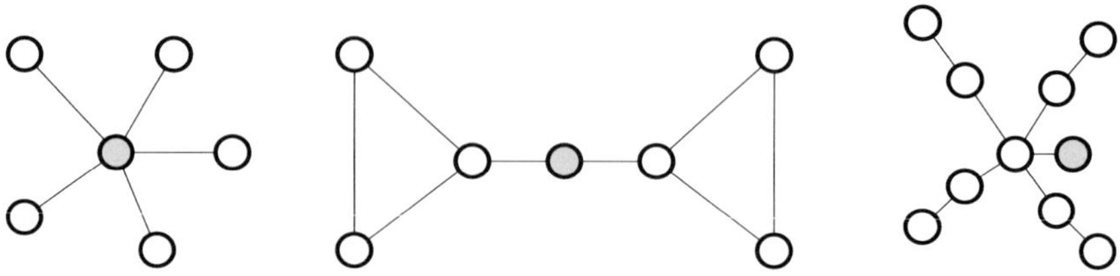

Figure 13: Left - Node with high Degree Centrality; Center - Node with high Betweenness Centrality; Right - Node with high Closeness Centrality

The distribution of degree centrality of nodes is an import network characteristic. Equal degree centralities of nodes lead to a homogenous network structure, but so called 'scale-free' (Barabási and Bonabeau 2003) networks have a hub and spoke structure with few highly connected nodes (hubs) and many little connected nodes (spokes) (Barabási 2015, p.29). In this case the degree distribution follows a power law.

In scale-free networks the same phenomena affect a system at many different scales. For example, an organization may be a scale-free network and the rules for forming teams apply at the personal, small-team, and large-team scale (Sheard 2007). A similar network characteristic is the so-called 'small world' network, which has high clustering of elements and a low path length between elements (Watts and Strogatz 1998). Path length refers to the number of elements one must pass to get from one random element to another random element; hence, path length is associated with closeness of elements. For example, a low path length in an organization allows fast communication (Sheard 2007).

Degree distribution is an indicator for network robustness. While a network stays usually intact when a little connected node breaks down, the whole network can fail in an directed attack on a highly connected node (Watts and Strogatz 1998). Many large networks are scale-free, e.g., the internet, and this characteristic has led to increased research. For example, Braha and Bar-Yam (2004) show that the connectedness of tasks in product development projects can follow a power law distribution. Figure 14 shows the structural differences and distributions for homogenous and scale-free networks.

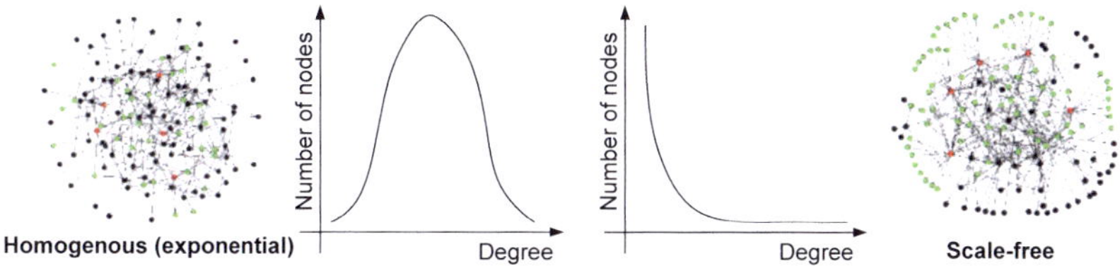

Figure 14: Homogenous and Scale-free Networks (based on Albert et al. (2000); Kreimeyer (2009))

2.2 Structural Aspects of the AEC Design System

After reviewing the theoretical foundations of systems science and related modeling techniques, this section addresses the real-world system "AEC Design". This section is structured into three parts: product, process, and organization of the AEC detailed design phase. Section 2.2.1 describes structural characteristics of the AEC design product. Section 2.2.2 focuses on the AEC design process, and section 2.2.3 describes the AEC design organization.

2.2.1 AEC Design Product

This section reviews structural characteristics of the AEC design product, which is the building design with a production process. AEC design products, here called buildings, often comprising a large number of different systems, usually have a long lifecycle compared to the design and production process. The design process often integrates a large number of different professions.

Product modeling usually distinguishes between different levels of detail for entities of the product structure. For example, Pimmler and Eppinger (1994) distinguish between subsystems and components. Relations between entities are often dependencies regarding heat, information, or electricity transfer. Also, spatial proximity can be a dependency.

Literature on product structures distinguishes between integrated and modular product structures. Modular structures include groups of entities, e.g., components that are highly connected within the group, but sparsely connected to the rest of the product structure (Baldwin and Clark 2000) (see table 1). Ulrich and Eppinger (2004, p.165) describe two attributes of modular product structures:

(1) Modules implement only one or few functions of the product.
(2) Interactions between modules are well defined.

In order to develop a modular product structure, it is important to define modules and set the relations between modules early in the design process. Modular product structures have both advantages and disadvantages when compared to integral product structures. Advantages are a parallelization of design of modules, economies of scale and higher innovation of technologies within modules (when modules are shared across different products), and flexibility for product adaptation (Mohamad et al. 2013). A high degree of modularity makes it possible to have "loosely coupled product creation organization in which each participating component development unit can function autonomously and concurrently" (Sanchez and Mahoney, 1996, p. 65). A disadvantage is a possible lower performance as compared to integral product structures (Ulrich and Eppinger 2004, p.166).

Mohamad et al. (2013) state regarding modular product structures in the AEC industry:

"Literature shows different uses of the term 'modularization' in the construction industry. Court (2009) defines modularity in production as an assembly system where modules consist of components that can be combined off-site and then delivered to the construction site. CII (2011) identifies potential improvements, such as lower cost, shorter schedule and better quality, through the use of pre-designed modules across several construction projects. Standardized modules can be combined to produce a customized product. Thus, the design phase becomes a configuration phase, in which designers combine available modules into a customized product (Jensen et al. 2009). Veenstra et al. (2006) introduce a platform-based methodology emphasizing the importance to balance standardization and variation in order to meet the different customer values. Lennartsson et al. (2008) emphasize the importance to balance customer value and delivery team value when defining product platforms and modules in industrial housing. The presented approaches apply modular design by using standardized modules across several projects."

Product Modeling needs modeling tools and accordingly the trend is to use Building Information Modeling (BIM) in the AEC industry. BIM uses an integrated database that all project participants can access with specific rights regarding what they can see and/or change (Both 2011; Eastman et al. 2008). BIM enables modeling product entities, e.g., components of the building, and relations between these components, e.g., spatial proximity, heat flow, airflow. Simulation tools can compute, e.g., building performance, code compliance, construction processes, and building costs. BIM can execute validity checks, e.g., regarding proximity of objects through identification of spatial conflicts between components, also called clash detection.

2.2.2 AEC Design Process

This section first reviews the design literature with a short digression into business process literature. Next, this section reviews the goals of process management, followed by a review of strategies for analyzing engineering design processes. Last, this section presents an overview of existing metrics for engineering design processes.

2.2.2.1 Characteristics of Design Processes

The terms 'engineering design' and 'product development' are nowadays used almost interchangeably, but this was not always the case. The terms stem from different schools of thought. Design had previously described the process of finding a solution to a well-defined set of requirements, while product development had previously described the overall process from collecting customer requirements through engineering design to production planning, conducted in an over-the wall manner (Motte et al. 2011). The "total design" (Pugh 1991, p.5f.) approach integrated the stages of the product development process (Motte et al. 2011) so that, in common terminology, design encompasses collection of customer requirements and considerations regarding product adaptation, production, and sales. This dissertation uses the term 'design process' in the sense of an integrated product development process through all stages.

Albers and Meboldt (2007) describe design as two concurrent processes: (1) learning about customer requirements ("system of objectives") and (2) finding ways to fulfill customer requirements by narrowing the design space ("object system") (figure 15). They describe the product development project system as follows:

"[...] product development can be described as the transfer from a system of objectives, being still vague at the beginning of the product development, to a concrete object system. I.e., the core activity of the product development is the continuous expansion and specification of a system of objectives, the creation of an efficient operation system and therefore the successful realization into an object system - the product" (Albers and Meboldt 2007).

The relationship between goal system and product system, as described by Albers and Meboldt (2007), assumes uncertainty regarding goals; this impacts all other domains of the product development system. Product related uncertainty (1) hinders exact identification of functions and components, because the final functions and components will only be identified throughout the design process, (2) hinders long-term process definition, because tasks and their dependencies can hardly be anticipated without knowing the functions desired by the customer, and (3) hinders pre-definition of organization structures and tools, because people's tasks are unknown. Hence, product-related uncertainty leads to a probabilistic and dynamic production system.

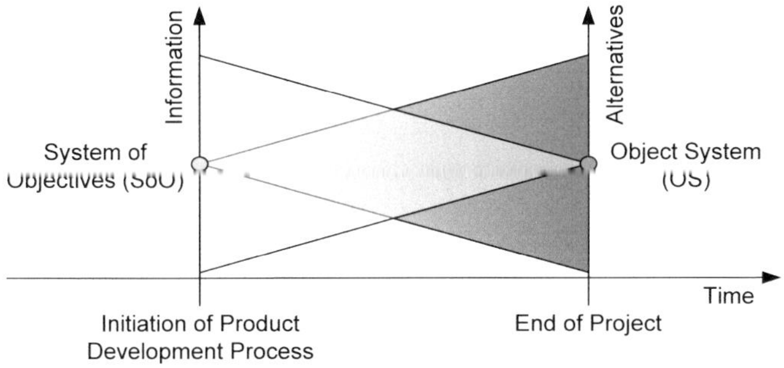

Figure 15: System of Objectives (SoO) - Object System (OS) in the Product Development Process (based on (Albers and Meboldt 2007))

Koskela and Kagioglou (2006a) describe, similarly to Albers and Meboldt (2007), two concurrent processes of (1) analysis and (2) synthesis. Design involves creativity and learning: designers apply creativity to develop solutions for unsolved problems. Users then review those solutions and in doing so they learn about their requirements, i.e., they extend their knowledge about their objectives. Next, designers refine and improve the prior solutions or develop completely new ones.

Hatchuel and Weil (2003) criticize Simon's (1996, p.132) description of design as problem solving, because Simon's description lacks the concept of creativity. That concept is: designers must generate new knowledge in order to solve a design problem. This new knowledge affects the design system: it can change the assumptions on which the requirements were based, thus changing the starting point for finding a solution and leading to an iterative cycle of analysis and synthesis. Accordingly, Rittel and Webber (1973) characterize design problems as "wicked", i.e., "ill-structured and pernicious" (Wynn and Clarkson 2005, p.35). Maier et al. (2011) describe the engineering design process as ill-defined, iterative, and complex. The wicked problem of design refers to uncertainty regarding requirements and constraints; creativity can help eliminate uncertainty by providing design solutions.

Design Methodology is:

"The study of how designers work and think; the establishment of appropriate structures for the design process; the development and application of new design methods, techniques, and procedures; and reflection on the nature and extent of design knowledge and its application to design problems" (Cross 1984, pp. vii-viii).

Waldron and Waldron (1996) distinguish between the process view and the artifact view on design methodology. They provide **definition one of process**:

"The design process can be viewed as a sequence of steps, such as clarification of the specifications and the environment in which the design will function, understanding the behavior, and establishing the operational constraints, including manufacture, servicing, marketability, usability, and disposability" (Waldron and Waldron 1996).

Hence, design can be regarded as a process with distinct entities, such as steps, tasks, or stages, with information flow relating them. Vajna (2005, p.371) compares business processes to engineering design processes (table 5), which are representative for processes in AEC design. This comparison highlights the complex and creative nature of design.

Table 5: Difference between Business Processes and Engineering Design Processes (Vajna 2005, p.371)

Business Process	**Engineering Design Process**
Processes are fixed, rigid, have to be reproducible and checkable to 100%	Processes are dynamic, creative, chaotic; many loops and go-tos
Results have to be predictable	Results are not always predictable
Material, technologies, and tools are physical (e.g., in manufacturing) and/or completely described (e.g., in controlling)	Objects, concepts, ideas, designs, approaches, trials (and errors) are virtual and not always precise
Possibility of disruptions is low, because objects and their respective environments are described precisely	Possibility of disruptions is high because of imperfect definitions and change requests
No need for dynamic reaction capability	There is definitive need for dynamic reaction capabilities

Uncertainty surfaces in the design process through iteration, i.e., the partial or complete repetition of an already completed task. Smith and Eppinger (1997) describe two categories of reasons for iteration:

(1) Repetition of an upstream task, because a downstream task discovers an error or failure to meet the upstream task's objectives.

(2) Repetition of a downstream task, because information coming from upstream is changed due a correction or change in goals.

Often though, interdependency of tasks and cyclic dependencies between tasks cause iteration. In this case, the design process begins with preliminary values and the process iterates until all task's objectives are met

(Ballard 2000b).[13] Faster and fewer iterations can reduce project duration and the DSM is an appropriate tool for modeling and analyzing task dependencies (Browning 1998).

Process Modeling is an important part of design process management, as it reveals the structure and dependencies of tasks and information flows. Wynn and Clarkson (2005) identify design process modeling as part of design methodology. They identify three dimensions in design process modeling (Wynn and Clarkson 2005, p.35):

- stage vs. activity-based models,
- problem vs. solution-oriented literature,
- abstract vs. analytical vs. procedural approaches.

This dissertation focuses on an analytical approach for process modeling, which consists of three steps:

(1) Decomposition of the overall design project into entities, such as phases and activities.
(2) Integration of entities based on information needs, i.e., finding information flow dependencies between tasks.
(3) Optimization of the resulting network regarding several factors, e.g., duration, cost, iteration, and risk.

Process models can be further classified based on whether they are (1) descriptive or (2) prescriptive (Wynn and Clarkson 2005):

(1) Descriptive models capture actual processes 'as-is' or describe typically followed procedures. Process mining (Aalst 2005, 2011) is a method for gathering data for modeling processes. This data

[13] Figure 7 provides an example of cyclic dependencies: the load of the structure - in this case the walls and roof - impact the size of the foundations but also the design of the structure itself. The structure carries the load, but the load is unknown unless the structure is designed. The size of the structure impacts design aspects of the overall building and a change in design (e.g., additional windows) impacts loads, and hence, structure.

usually stems from an IT-system, where interactions between people leave traces, e.g., in the form of logs. Process mining discovers and collects data that can describe actual interactions between people. Process mining has been applied to business process management.

(2) Prescriptive models aim at improving performance and target a specific group of people and/or class of design problems, e.g., mechanical engineering design or AEC design. Prescriptive models provide a 'should' perspective; they tell designers what to do. Successful implementation of the prescriptive model relies on (1) valid understanding of the prescriptive model and (2) the fit between the prescriptive model, which had been defined in advance, and the actually conducted process (Eckert and Stacey 2010).

Process models capture dependencies between tasks, and one important gap between model and reality is information processing within each task. Browning et al. (2006) highlight the importance of knowledge in the design process: people use their knowledge to conduct creative tasks, which may create new knowledge. Decoding information to knowledge and encoding knowledge to information depends also on a person's constructed reality, i.e., his/her mental model (Browning et al. 2006).

2.2.2.2 Goals of Design Process Management

Browning et al. (2006) argue that the design process shall be regarded as a system and that the design system can be 'engineered' to improve project planning and organizational learning. Following the definition presented in section 2.1.2, the purpose of a system guides development of its structure. Hammer and Champy's (1999, p.35) definition of process defines value delivery as a purpose of a process. **Definition two of process** is:

"We define a business process as a collection of activities that takes one or more kinds of input and creates an output that is of value to the customer." Hammer and Champy's (1999, p.35)

Koskela (2000, p.27) provides a more detailed perspective on the goals of production system management. He details value delivery by separating it into three parts.

(1) Providing the product,
(2) Minimizing waste,

Maximizing value.Koskela (2000, p.27) describes three generic tasks of management to achieve the described goals: planning, execution, and controlling. A continuum of different approaches to conduct these three tasks exists. Koskela and Howell (2002) present two typical approaches to these three tasks of management: (1) traditional project management and (2) Lean Construction[14].

Table 6 compares traditional project management to Lean Construction. Management-as-planning refers to central planning and then giving orders to execute the plan (production). The focus of management-as-planning lays on the planning part of management. In contrast, management-as-organizing focuses on enabling decentral sub-units to interact with each other. Then, management focuses on structuring the setting so that interaction between sub-units leads to desired outcomes.

Classical communication theory refers to transmission of information. In the case of traditional project management, an order is communicated. In contrast in Lean Construction, the Language-Action-Perspective (Flores 1981, p.78) refers to the process of making requests, coordinating requirements, and making commitments.

The thermostat model refers to comparing process output to planned performance. In case both values differ more than the allowed range, the

[14] Section 2.3.2.2 contains a description of Lean Construction.

thermostat model takes corrective action so planned performance can be reached. In contrast, the scientific experimentation model refers to documenting a standard process, stating a hypothesis regarding performance, and evaluating the hypothesis by conducting an experiment, i.e., executing the process. Hypothesis testing leads to quick improvement cycles of the process (Koskela and Howell 2002).

Table 6: Approaches to Management (based on Koskela and Howell (2002))

Task of Project Management	**Traditional Project Management**	**Lean Construction**
Planning	Management-as-planning	Management-as-organizing
Execution	Classical communication theory	Language Action Perspective
Controlling	Thermostat Model	Scientific Experimentation Model

The following sections present the three tasks of project management - planning, execution, and controlling - in further detail based on the Lean Construction approach to project management.

2.2.2.3 Planning

In the Lean Construction approach to Project Management, planning includes production system design. Ballard et al. (2001a) provide a number of means to increase value generation and reduce waste through work structuring. Work structuring is part of production system design, and it consists of decomposition, integration, and optimization. Based on Ballard (1999) Tsao et al. (2004) describe work structuring by six questions:

(1) In what units will work be assigned to groups of workers?
(2) How will work be sequenced?
(3) How will work be released from one group of workers to the next?
(4) Will consecutive groups of workers execute work in a continuous flow process or will work be decoupled?
(5) Where will decoupling buffers be needed and how should they be sized?
(6) When will different units of work be done?

2.2.2.4 Execution

During production system operation designers execute tasks in order to generate information. Critical to the effective generation of information is coordination between activities. Malone and Crowston (1990) structure coordination into four processes with components (table 7):

Table 7: Processes underlying Coordination (Malone and Crowston 1990)

Process Level	**Components**	**Examples of Generic Processes**
Coordination	Goals, activities, actors, resources, interdependencies	Identifying goals, ordering activities, assigning activities to actors, allocating resources, synchronizing activities
Group decision-making	Goals, actors, alternatives, evaluations, choices	Proposing alternatives, evaluating alternatives, making choices (e.g., by authority, consensus, voting)
Communication	Senders, receivers, messages, languages	Establishing common languages, selecting receiver (routing), transporting message (delivering)
Perception of common objects	Actors, objects	Seeing same physical objects, accessing shared databases

Proponents of Lean Construction apply the Last Planner System (LPS) (Ballard 1994, 2000c) to coordinate production processes (see section 2.3.2.2 for a description of LPS). Pall (2000) presents a similar approach to process coordination. In the field of Systems Engineering Pall's (2000) network of commitments has received attention as a method for process coordination (e.g., Browning et al. (2006); Browning and Ramasesh (2007)); it also includes practices for planning and improvement. Both methods, the LPS (Ballard 1994, 2000c) and (Pall 2000)'s network of commitments approach overlap in several aspects: both advocate pull planning, process coordination based on Flores' (1981) LAP[15], and measuring process reliability.

2.2.2.5 Controlling

Improvement of production systems often leads to structural change or adaptation. Production Systems are socio-technical systems, and in a social context improvement relates to learning. Looking at the production system structure, this dissertation focuses on organizational learning. According to (Dodgson 1993, p.377) organizational learning

"can be described as the ways firms build, supplement, and organize knowledge and routines around their activities and within their cultures, and adapt and develop organizational efficiency by improving the use of the broad skills of their workforces."

From a systems perspective, organizational learning relies on the GST principle (see section 2.1.4.1) of self-reference. Probst and Büchel (1997, pp.35ff.) describe three types of learning in organizations (Probst and Büchel 1997, p.35).

(1) Single-loop learning is triggered by a deviation between results and prior established goals. Learning consists of an adjustment of behavior in order to achieve planned goals.
(2) Double-loop learning questions existing goals of the organization and can result in the change of goals and related structures and possible behaviors.
(3) Deutero learning focuses on the process of learning, i.e., on learning how the organization learns. Learning proceeds through reflection of results, problem solving strategies, and learning procedures.

[15] Ballard's (1994, 2000c) original description of the LPS does not mention LAP, but it was later added by a series of papers (Howell et al. 2004; Macomber et al. 2005; Macomber and Howell 2003).

In order to exploit its full learning potential, an organization must implement all three feedback loops and provide flexibility in behavior, goals, and learning processes. Flexibility enables adaptation. Figure 16 visualizes the relation between these three types of learning and shows the feedback loop from results to different parts of the organization. These feedback loops implement the principle of self-reference.

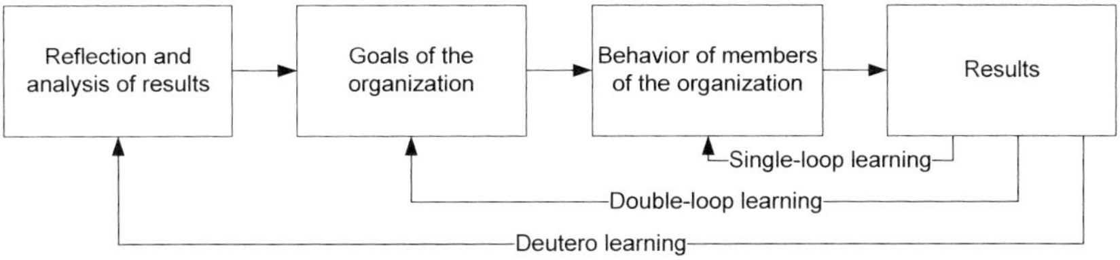

Figure 16: Three types of Learning in Organizations ((Probst and Büchel 1997, pp.35ff.) based on Argyris and Schön (1978))

Learning requires transparency of results, because only transparency of results enables comparison to goals, and in turn questioning of goals and reflection. Measurement of results is a first step for learning to change behavior. In engineering design, measurement of results usually focuses on controlling of performance, i.e., time, cost, and quality. Ballard (2000c) adds measurement of process reliability with the PPC value (see section 2.3.2.2).

Learning is a continuous process, which repeats itself. Shewhart (1939) and Deming (2000, p.88) explain continuous improvement with the Plan-Do-Check-Act (PDCA) cycle. The PDCA is a model for continuous improvement and it is rooted in the scientific method (see section 2.3.2.1).

2.2.3 AEC Design Organization

Organization theory deals with the problem of dividing a large task into chunks that are manageable by people or teams. Two general definitions for 'organization' exist (Reichwald and Möslein 1997, p.2):

(1) Organization as an instrument refers to the sum of means to achieve a goal.
(2) Organization as an institution refers to a social construct with a goal and a structure.

This dissertation defines 'organization' based on the institution view. A large body of literature regarding organization theory exists that shows several streams of organization theory (Reichwald and Möslein 1997, p.6). The works of Lawrence et al. (1967) and Thompson (2010) are part of the systems-theoretical stream of organization theory, and this dissertation adopts a systems-theoretical perspective to organizations. The systems-theoretical stream regards organizations as open- and self-organizing systems (Reichwald and Möslein 1997, p.6).

The following section is structured as follows: the first part describes organization design, followed by a description of formal organization, and descriptions of communication and informal organization. The section closes with a description of organization development.

2.2.3.1 Organization Design

The main function of organization design in design projects is partitioning and integration of the overall project task (Sosa and Mihm 2008, p.165). Partitioning and integration refers to dividing a task into subtasks, assigning these subtasks to people or teams, and then integrating people or teams. Division of the organization into a modular structure comes along with integrative mechanisms that span module boundaries (Lawrence and Lorsch 1967). Galbraith (1974) identifies information as what is processed in organizations: thus, organization design is closely related to structuring communication between people in the organization, because communication transports information.

Different authors use the terms 'integration' and 'coordination' interchangeably. In this dissertation 'integration' refers to establishing connecting points or 'bridges' between entities of the organization, e.g., people or teams. 'Coordination' refers to the definition and sequencing of tasks, assigning them to people, allocation of appropriate resources for completing tasks, and synchronizing tasks during execution (Malone and Crowston 1990). The purpose of integration is to ease coordination. Sherman (2004) highlights the importance of coordination: even when levels of integration are appropriate, integration alone is not sufficient to avoid coordination problems.

Organization contingency theory researches the dependence of organization design on other project attributes. According to Lawrence and Lorsch (1967) organization contingency theory is the application of systems theory to organizations. Organization contingency theory postulates that organizations must be fitted to circumstances of the enterprise in order to be efficient, i.e., there is no "one size fits all" approach to organization design. Several attributes impact success of organization design:

- Project goals (Burns and Stalker 1961)
- Project Environment (Burns and Stalker 1961)
- Information processing dependencies (Thompson 2010)
- Uncertainty (Tushman and Nadler 1978)
- Organization size (Pugh et al. 1969)
- Technology (Burns and Stalker 1961)

Organization contingency theory has been applied to project management. Sauser et al. (2009) and Shenhar and Dvir (1996) argue that each project differs in its characteristics from others and that critical success factors are not the same and not generally applicable to all projects. Engwall (2003) criticizes that project management theory lacks focus on projects' environment: it is necessary to regard a project as interconnected with its environment in order to gain a correct understanding of the project itself.

2.2.3.2 Formal Organization

Formal organization refers to an organization's structure and procedures. Organization design establishes the formal organization and this includes, but is not limited to, lines of authority, reporting relations, behavior required according to organizational rules, patterns of decision making, patterns of communication, incentive structures, and problem solving approaches (Donaldson 1999; Sosa and Mihm 2008). Henderson and Clark (1990) present four elements of formal organization: workgroups, communication channels, information filters, and a repertory of problem solving strategies.

Organization architecture is a subset of the formal organization. Eppinger and Browning (2012, p.81) and Ulrich and Eppinger (2004, p. 23) distinguish between two types of relations in organization architectures: (1) reporting relations, which are mostly vertically arranged, and (2) lateral relations, which are mostly horizontally arranged. Their definition of organization architecture focuses on information flow. Eppinger and Browning (2012, p.80f.) describe lateral relations as an "interaction network", where interaction refers to information flow between units of the organization. These interactions can be "formal or informal peer-to-peer communications" through different communication channels and interactions "based on relationships of authority, responsibility, accountability, contractual obligations, and so on" (Eppinger and Browning 2012, p.80).

Partitioning of organizations builds artificial boundaries between groups of people to establish internal focus within each group. Partitioning groups people with strong work-related relations. Two basic types of formal organizations exist (Sosa and Mihm 2008):

(1) Functional organization: boundaries exist between company functions, i.e., people are grouped by discipline.
(2) Project organization: boundaries exist between projects, i.e., people from several disciplines are grouped by project.

Matrix organization is a third type of formal organization, combining characteristics of functional and project organization to form cross-functional teams. Cross-functional teams combine experts from several disciplines. Experts stay connected with their functional group but are at the same time responsible for project success. Thus, two lines of reporting are defined: (1) to the functional leader and (2) to the project leader.

Galbraith (1971) describes the space between functional and project organizations as a continuum. Sosa and Mihm (2008) place organization structures in the context of market change and knowledge change; both are related to uncertainty in the environment. Project organizations perform better when markets change quickly and specialists' knowledge changes slowly, because interdisciplinary teams can collaboratively develop new products. Functional organizations perform better when market change is slow and knowledge change is quick, because people from the same discipline can better exchange knowledge from their discipline. Matrix organizations cover situations where both, market and knowledge change, are relevant for the organization. Two typical types of matrix organizations exist:

(1) Light weight matrix organization: members are mainly associated with their functions and members do not report to the light-weight project manager. Functional members are responsible for staffing decisions and budgets (Ulrich and Eppinger 2004, p.26).

(2) Heavy weight matrix organization: members are mainly associated with the project, and the heavyweight project manager has budget authority and is involved in performance evaluation of team members (Ulrich and Eppinger 2004, p.26).

While the types of organizations as described look at the placement or location of organizational boundaries, Worren (2012, p.168) analyzes the 'height' of boundaries between parts of the formal organization based on work level interdependencies (figure 17, left hand side). He defines work-level interdependency as a combination of uncertainty in and importance of information to be exchanged (figure 17, right hand side) (Worren 2012, p.201). Tushman and Nadler (1978) argue that the level of integration depends on the amount of uncertainty: higher uncertainty demands stronger integration.

Work level interdependencies are chaotic when both importance and uncertainty are high. In this case an integrated organization provides an efficient design. In contrast, work-level interdependencies are well-documented, predictable and affect few outcomes, when uncertainty and importance are both low. In this case a partitioned formal organization provides an efficient design.

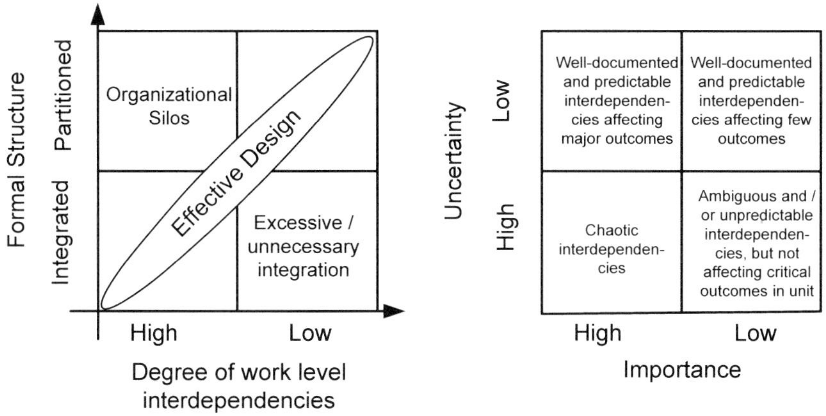

Figure 17: Relationship between Formal Structure (grouping) and Work Process Interdependencies (Worren 2012, p.168); Framework for characterizing Degree of Interdependency between Organizational Sub-units (Worren 2012, p.201)

Aside from people being members of the same group, e.g., a functional or project group, several other integrative mechanisms exist. Browning (2009) identifies 15 integrative mechanisms for integration of groups in a multi-team environment. Some integrative mechanisms also apply to single-team environments. Table 8 presents representative integrative mechanisms.

Table 8: Representative Integrative Mechanisms

Integrative mechanisms (Browning 2009)
Improved information and communication technologies – collaborative tools, linked computer-aided design (CAD)/ computer-aided engineering (CAM) systems, email distribution lists, tele- and video conferencing, common databases (easily accessed and shared), and so on.
Training – especially in team building (and "system team building" and "program building"); raising awareness about integration needs and roles.
Collocation – physical adjacency different teams and/or organization members.
Traditional meetings – face-to-face gatherings for information sharing and/or decision making.
"Town meetings" – not to share technical information but to boost camaraderie, increase awareness of program-wide issues, and a greater shared culture.
Manager mediation – "up-over-down" (hierarchical) issue mediation schemes; heavyweight product managers; orchestrators; and integrators, including supply-chain integrators.
Participant mediation – boundary spanners, liaisons, engineering liaisons, and conflict resolution engineers.
Interface management groups – Integration teams tasked with ensuring ongoing or incident-specific mediation of interface issues.
Standard processes (that include specified deliverables or work products) – shared routines and procedures; explicit delineation of interface characteristics and metrics for evaluating interface effectiveness; includes interface contracts and scorecards.
"Boundary objects" – objects operated on by those on both sides of an interface, such as shared models and repositories.
Incentive systems – rewards and/or penalties for work performance in relation to interfaces or other teams.
Shared interpretations of design problems.
Shared knowledge.
Shared ontologies – common terminology across teams for products, processes, and tools.
Situation visibility – shared visual orientation of a team's activities and results in "the big picture".

In the field of product development, Browning (2009) presents a six step approach to designing organization architecture in a multi-team environment (figure 18). The first step in designing organization architectures is to understand the architectures of products and processes, to which teams are being assigned in the second step. The third step groups teams based on their interdependencies, and step four integrates teams through integrative mechanisms. During work execution, step five manages interfaces. Step six re-assesses and, if necessary, executes preceding steps in order to adapt the organization architecture.

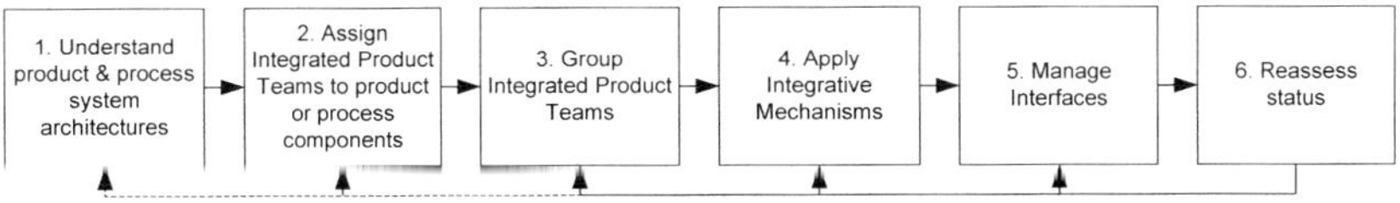

Figure 18: A Design for Integration (DFI) Process (Browning 2009)

Literature from the field of SNA also researches organization architecture and integration. Hansen (1999) applies Granovetter's (1973) distinction between strong and weak relations between people (here called ties) to analyze potential for knowledge sharing. Hansen (1999) argues that weak relations, i.e., infrequent and distant, are efficient for interactions between organizational subunits, if the knowledge to be transferred is not complex. He recommends strong relations, i.e., close and frequent, for the transfer of complex knowledge. Levin and Cross (2004) describe the importance of trust in others' competence and benevolence when establishing strong and weak ties.

2.2.3.3 Communication and the Informal Organization

Organization design establishes the formal organization, and within the boundaries of the formal organization, the informal organization develops. Birrell (1981) describes the importance of the informal organization for managing construction processes. From the information processing perspective (Galbraith 1974), people communicate to transfer information. Information is the material flowing between work-stations of the design process, and work-stations transform input information to output information by adding value (in the form of additional information). Information flow in design differs from material flow in production in at least three ways:

(1) The 'matter' of designers is information. While material flow in 'making' is mostly visible, information flow in 'designing' can be invisible. This makes it harder to trace the actual flow of information.
(2) Complexity hinders the identification of waste in design, and it is often the case that necessary vs. non-value adding tasks can be differentiated only after the design has been completed (e.g., Browning 2003).
(3) Iteration in 'making' represents waste, whereas iteration in 'designing' may offer an opportunity for designers to deepen their understanding of the task and explore alternatives, so that they can deliver an outcome of greater value to the customer. Value-adding iteration is to be encouraged. Iteration is called wasteful, if it can be eliminated from the process without a loss of value or risking the success of the project; this so-called negative iteration (Ballard 2000) should be avoided.

Information flow and communication differ in their characteristics. Information flows between tasks, while communication connects people. Hence, communication is a means for coordination between people, who conduct tasks (Maier et al. 2008). Koskela and Howell (2002) compare two models of communication:

- Model 1: Classical Communication Theory (Shannon and Weaver 1959):

Information flows originate in a source. A transmitter encodes the signal which then flows through a channel to a receiver, which decodes information for the destination. Shannon and Weaver (1959) point at three sets of problems with communication based on this model: (1) accuracy of transmission in the communication channel, (2) accuracy of meaning through decoding and encoding, and (3) effectiveness of communication in regards of change of behavior at destination. The model does not take into account that people may interpret information differently, even if the meaning has been accurately encoded, transmitted, and decoded. Koskela and Howell (2002) criticize this model in the context of management, because of its one-way communication. They refer to this type of communication as "dispatching" or execution of an order.

- Model 2: Language Action Protocol (Flores 1981)

Flores (1981, pp.77f.) presents the "Language Action Protocol" (LAP) as a model of communication. LAP consists of generic speech acts which in combination result in communication between people. Flores' communication theory is rooted in the idea of constructivism, and LAP aims at aligning people's constructs of reality to achieve successful communication. Flores' LAP assumes communication as conversations, and this term highlights two-way communication between people. Conversations develop as a cycle between a customer and a supplier. People align their constructed realities through conversation, which establishes a feedback cycle.

Macomber and Howell (2003) adapt Flores (1981, p.78) LAP for Lean Project Management. Conversations between customer and provider consist of four steps: request, commitment, declaration of completeness, and declaration of satisfaction (figure 19). This structured communication cycle includes coordination between customer and supplier through two-way communication regarding requirements of the request and fulfillment of conditions of satisfaction. Application of Flores' coordination cycle in an organization establishes a network of commitments.

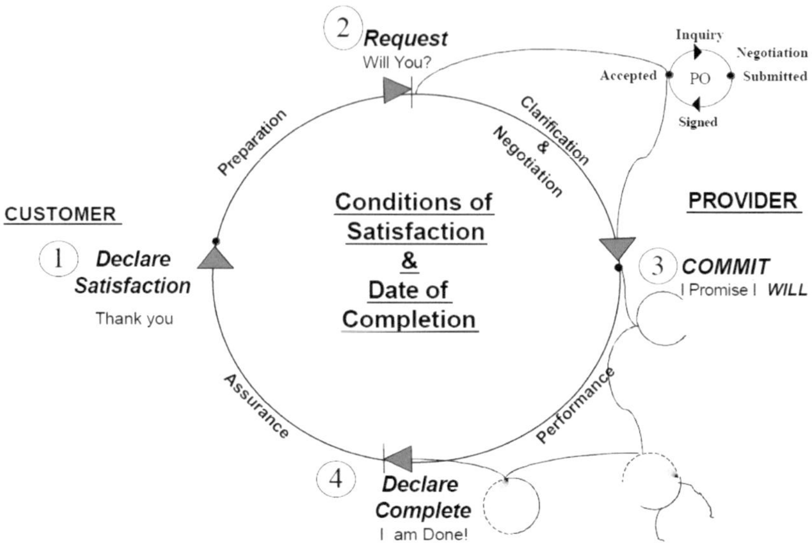

Figure 19: The Conversations for Action (Macomber and Howell 2003)

Different types of communication structures can emerge in informal organizations. Both (2006, p. 280f.) presents typical structures for communication between people. She recommends a network structure with direct connections between people to increase organizational flexibility (as compared to a star-shaped structure which includes a strong information hub). Further, she highlights the importance of mechanisms for access to and distribution of information to shape a network structure with direct connections. Allen (1977) researches the impact of physical distance on face-to face communication between people. He finds that increased distance lowers the probability of communication between people. Sosa et al. (2002) show the relation between the choice of communication technology (face-to-face, telephone, email) and physical distance between people.

Maier et al. (2008) research correlations between factors influencing communication in product development, and they identify the following core factors that influence communication: mutual trust, collaboration, roles and responsibilities, availability of information about product specifications, handling of technical conflicts, 'do you know what information the other party needs', autonomy of task execution, and overview of sequence of tasks in the design process.

Priven and Sacks (2013) show that implementation of the Last Planner System ™ (LPS)[16] can lead to a network structure with direct connections, and that implementation of the LPS strengthens ties between construction crew members with different backgrounds, e.g., between members of different trades.

Efficiency and effectiveness are characteristics of communication. Chinowsky et al. (2008) highlight the importance of communication as an enabler for trust between people, which is necessary to achieve high performance teams. Eckert et al. (2001) explain the need for targeted communication among members of a design team in order to avoid information overload. Eppler and Mengis (2003) visualize the problem of information overload (figure 20): people's decisions become more accurate with increasing amounts of available information until they suffer from information overload. Mihm et al. (2003) argue that people will then cut some communication, which increases the risk of missing important information and which in turn might lead to wasteful rework. Eppler and Mengis (2003) identify several causes of information overload and group them in five categories: personal factors, information characteristics, task and process parameters, organization design, and information technology.

[16] Section 2.3.2.2 describes the LPS.

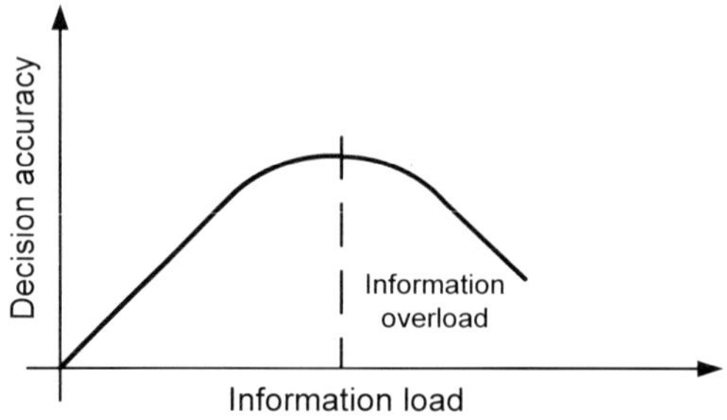

Figure 20: Information Overload as the inverted U-Curve (Eppler and Mengis 2003)

Nonaka (1990) describes the positive effects of excess communication on design team creativity. Kratzer et al. (2008) analyze misalignments between formal and informal organization architecture (figure 21), and they find that additional communication (defined as the difference between actual communication in the informal organization and planned communication through formal organization) between people can increase creativity. At the same time, additional communication reduces time efficiency. Thus, a conflict exists between efficiency and effectiveness in organization design: increased communication between people may lead to increased effectiveness by fostering higher creativity, and thus possibly to better delivery of customer value. At the same time, increased communication reduces efficiency, because sifting through unneeded information takes time away which could be used for other productive activities.

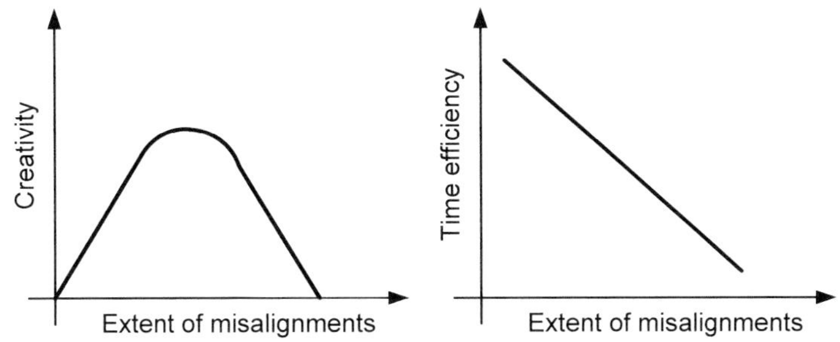

Figure 21: Extent of Misalignments and its Impact on Creativity, Time Efficiency

To summarize, structure, effectiveness, and efficiency of a communication network can be influenced through organization design. The structure of the communication network influences, among others, effectiveness and efficiency by directing communication between people. A conflict exists between efficiency and effectiveness in communication: under otherwise equal circumstances, increased communication may lead to increased effectiveness, but at the same time it reduces efficiency. When designers suffer from information overload, increased communication reduces both efficiency and effectiveness.

2.2.3.4 Organization Development

Organization design sets the characteristics of an organization, but the organization is subject to change. Change can originate in several sources. Organization contingency theory describes organizations as open systems that are subject to change in the environment.

Sosa and Mihm (2008) describe organization development in New Product Development (NPD): "Product development is a dynamic process that goes through very distinct phases. Yet, research has paid very little attention to the dynamics of organizations within NPD [New product Development] projects. How do formal and informal organizations differ (or should differ) across project phases? As projects progress and the informal organization evolves, should the formal organization adapt?"

Tsao et al. (2004) show that implementation of innovative solutions demands change of procedures across firm boundaries. Sheffer (2011, pp. 98f.) describes that innovations in the AEC industry can change established standards of building design and construction procedures. Implementation of these innovations demands companies to change.

Implementation of Lean Management (see section 2.3.2) aims at continuous improvement, which can affect organization architecture. Spear and Bowen (1999) describe the relationship between rigidity and flexibility in the Toyota Production System (TPS). The rigid rules of the TPS enable flexibility through learning and improvement.

Dooley (1997) describes design principles for complex adaptive organizations (table 9).

Table 9: Design Principles for Complex Adaptive Organization (based on Dooley (1997)

Create a shared purpose
Cultivate inquiry, learning, experimentation, and divergent thinking
Enhance external and internal interconnections via communication and technology
Instill rapid feedback-loops for self-reference and self-control
Cultivate diversity, specialization, differentiation and integration
Create shared values and principles of action
Make explicit a few but essential structural and behavioral boundaries

Pulm (2004, p.121) criticizes that research in organization theory focuses on static organizations which change from time to time in a top-down manner. He proposes that organization shall emerge instead. He focuses on the emergence of teams within an organization and presents five characteristics of an organization (table 10) that supports emergence of teams. This approach to organization development demands autonomy through a high degree of individual responsibility and a low degree of rigid organizational hierarchies and structures (Pulm 2004, p.123).

Table 10: Emergence of Teams in Organizations (Pulm 2004, p.123)

System emergence	Characteristics of organizations
Definition of boundaries	Definition of scope, definition of people involved in the team, definition of timeframe, comparison with other existing teams.
Generation of resources	Definition of goals and tasks (responsibilities, tasks, decisions), and definition of schedule for tasks.
Structuring	Integration with other teams regarding responsibilities, definition of interface with other teams, structuring of team internal tasks, etc.
Process control	Definition of means of communication, within the teams and with other teams (content, schedule, media for communication); observation of team development.
Reflection	Reflection what the goal of the team is/ was and whether the goal was reached. Reflection whether team composition is/ was appropriate.
Genesis	Based on team results the team defines new tasks or emerges into new teams.

To summarize, influences from within the organization and from its environment can make adaptation of the organization necessary. Organizations can change through top-down decision or they can adapt through emergence. Reflection and genesis implement feedback loops that support self-organization and emergence of the system.

2.3 A Production System Perspective on Design

This section addresses the purpose of the AEC design system. This dissertation regards AEC design as a production system. From this position, design is a process which produces a 'recipe' for assembly of a product. This recipe includes a description of the product as well as of the assembly process.

The term 'production system' originated in the stationary industry and a production system encompasses means of production, e.g., production facilities, machinery, and labor, materials and (semi-) finished goods, and the rules and methods which govern production (Kirsch 2009, p.14). Kirsch (2009, p.15) characterizes production systems as (see table 1 for a description of the following properties):

- socio-technical, because production systems include technical and social elements,
- complex, because relations between elements are dynamic, i.e., they change over time. Change is often desired, for example through improvement processes,
- open, because production systems interact with the systems environment,
- goal-oriented, because they generally aim at designing production processes which are aligned with the goals of the overall production endeavor,
- probabilistic, because internal uncertainty (lack of knowledge regarding the production system itself, e.g., quality of soil in earthworks) and external uncertainty (lack of knowledge regarding the environment, e.g., unpredictable weather conditions) make production system behavior hard to predict.

Projects establish a production system that often connects several companies involved in a project: the production system includes means of production of different companies and establishes rules and methods across different companies (Ballard et al. 2001b). Means of production and (semi-)finished products constitute elements of the production system. The rules and methods of production management and influences from the context establish relations between the elements of the production system.

Ballard and Koskela (1998), Huovila et al. (1997) and Koskela (2000) regard design as a production system and apply the Transformation-Flow-Value (TFV) theory to design management. Koskela (2000, p.111) describes differences between construction production and production in design:

- "There is much more iteration in design than in physical production.
- There is much more uncertainty in design than in production.
- Design is a non-repetitive (i.e., a project type) activity, production is often repetitive."

2.3.1 Transformation - Flow - Value Theory of Production

Koskela (1992, 2000) developed the TFV theory of production, which provides a theoretical basis for production system design in construction.

Koskela (2000, p.21) slices approaches for production management into three layers as shown in figure 22: concepts, principles, and methodologies. The conceptual layer answers the fundamental question "what is production?" (Koskela 2000, p.21). The principles layer explains relationships between different concepts. The methodologies layer consists of "methods, tools, practices, etc." (Koskela 2000, p.21) that follow concepts and principles. Koskela (2000, p.21) describes a theory as consisting of concepts and principles.

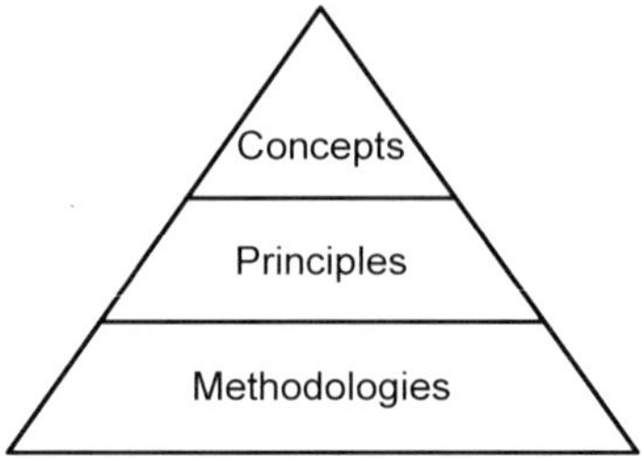

Figure 22: Relationship between Concepts, Principles, and Methodologies (Koskela 2000, p.21)

Integral to Koskela's (1992, 2000) TFV theory are three competing perspectives on production management: transformation, flow, and value. The TFV theory explains production management as finding a balance of these three perspectives that is aligned with the goals and environment of the production system. The three perspectives are explained next.

2.3.1.1 The Transformation Perspective on Production

The transformation perspective describes production as "a transformation of inputs to outputs" (Koskela 2000, p.89). According to the transformation perspective the production process consists of a series of activities, which generate the product. Management focuses on the proper execution of tasks and on responsibilities for tasks. The Work-Breakdown-Structure (WBS) can help in partitioning a production process into tasks, often through several hierarchical layers. Then, responsibility, budgets, and durations are assigned to each task prior to execution. Often the Critical Path Method (CPM) serves a tool for ordering the sequence of tasks. During execution, Earned Value Management (EVM) is often applied to measure the progress of each task. Figure 23 shows exemplarily the hierarchical decomposition of a production process into tasks (here called subprocesses).

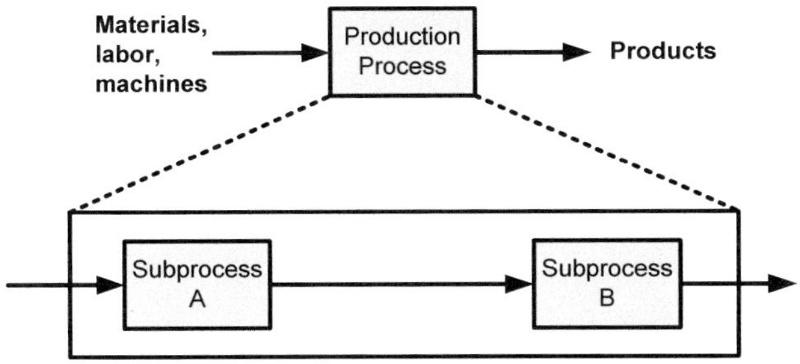

Figure 23: Transformation Perspective on Production (Koskela 2000, p.42)

Koskela (2000, p.254) criticizes production management for its concentration on the transformation perspective. This concentration can lead to local optimization of tasks, which causes inefficiencies for the overall production system. Ballard and Koskela (1998) criticize that design management often neglects the integration of tasks that are interrelated, either due to dependencies regarding the flow of information, or because they fulfill the same customer requirement. Koskela et al. (2002) state: "the transformation view is instrumental is discovering which tasks are needed. In a production undertaking and in getting them realized, however, it is not especially helpful in figuring out how to avoid wasting resources or how to ensure that customer requirements are met in the best possible manner."

2.3.1.2 The Flow Perspective on Production

According to the flow perspective a task does not only entail transformation (processing), but also the generic sub-tasks of moving, waiting, and inspection (Koskela et al. 1997). This extension of the process model links tasks as shown in figure 24. Of the four sub-tasks only processing is value adding, while the tasks of moving, waiting, and inspection are either necessary or wasteful.

Information is what primarily flows through the design process (Browning 2001; Eppinger 2001). For instance, 'building designer A' processes incoming information, e.g., building plans, and adds value by enhancing information, e.g., adding components of building system 'plumbing'. This information is then inspected and following a successful inspection, moved to 'designer B'. There, information waits for processing, e.g., by adding components of building system electrical to the plans. Figure 24 shows the process flow of two tasks in which non-value adding sub-tasks are tinted grey.

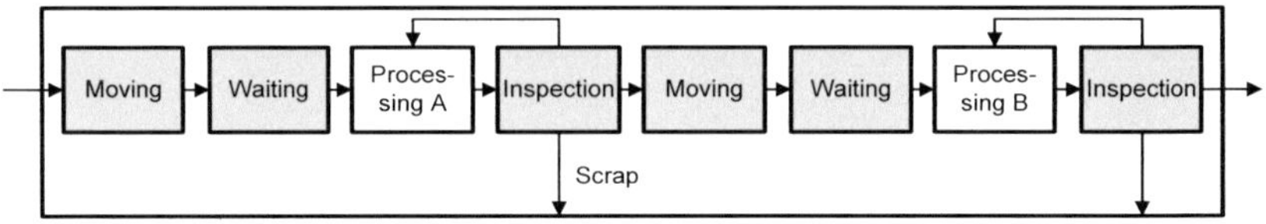

Figure 24: Flow Perspective on Production (Koskela 2000, p.56)

The flow perspective describes that a local optimization of tasks does not necessarily lead to a global optimum. Instead, the flow perspective demands optimization of tasks in the context of the overall process. It aims at achieving a reliable and steady workflow through elimination of root-causes for variation.

Methods supporting the flow perspective are value stream mapping (VSM) (Rother and Shook 2003) and, specifically in design, process DSM (Browning 2001; Steward 1981). Design processes often include interrelated tasks, i.e., the tasks are subject to iteration because they are connected by feedback loops. Iteration in design is not per se wasteful, it can also add value during the design process (Ballard 2000b). McManus (2005) combine DSM and VSM for design processes.

2.3.1.3 The Value Perspective on Production

According to the value perspective each task contributes to the delivery of customer value. The term customer does not only refer to the end customer of the product, rather all following tasks are also considered customers with their specific requirements.

Tuholski (2008, p.39) distinguishes between three objectives of AEC design: "(1) design of the facility, (2) design of the building design process, and (3) design of the building construction and supply chain processes." Each objective is client centric but achieves attributes of customer value in different ways. The building design process focuses on the fulfillment of requirements and expectations, thereby generating customer value in the form of facility design (Tuholski 2008, p.40). The design of the building design process is important, because it can affect the quality of the resulting facility design (Simon 1996, p.150). The building construction and supply chain processes realizes the facility design and fulfills customer requirements regarding facility delivery time, cost, quality, and other possible expectations (Tuholski 2008, p.40). Figure 25 shows the relationship between supplier and customer: for successful value delivery the supplier must collect, understand, fulfill the customer's requirements and expectations.

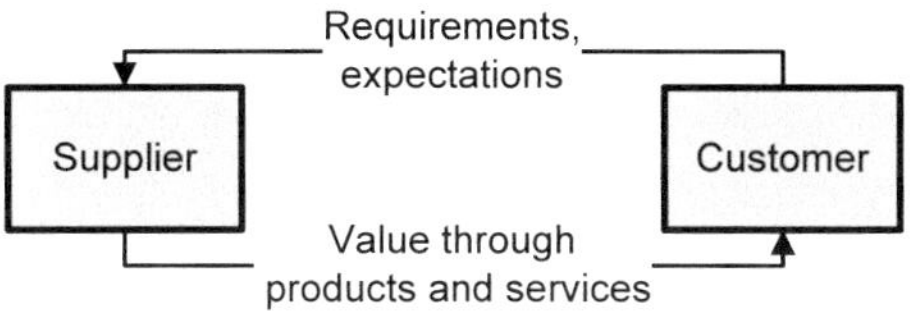

Figure 25: Value Perspective on Production (Koskela 2000, p.75)

Koskela (2000, p.120) names Quality Function Deployment (QFD) and value engineering as methodologies that support the value perspective.

2.3.2 Lean Management as a Concept for Managing Production Systems

2.3.2.1 Toyota Production System and Lean Production

Toyota developed a production system from the 1950s which served as the blueprint for the so-called 'Lean Production'. This term was coined in the seminal study conducted by Womack et al. (1990), which documented Toyota's approach to managing its production system. Several other scholars provided descriptions of the Toyota Production System (TPS) (Liker 2004; Ohno 1988; Shingo 1989; Spear and Bowen 1999). The term 'lean' refers to small amounts of materials and (semi-) finished goods in the production system and also to the core concept of reducing waste in the production system by improving it continuously.

Shah and Ward (2007) define the lean production system as follows: "Lean production is an integrated socio-technical system whose main objective is to eliminate waste by concurrently reducing or minimizing supplier, customer, and internal variability".

Spear and Bowen (1999) identify four rules which underlie the TPS:

(1) "All work shall be highly specified as to content, sequence, timing, and outcome.
(2) Every customer-supplier connection must be direct, and there must be an unambiguous yes-or-no way to send requests and receive responses.
(3) The pathway for every product and service must be simple and direct.
(4) Any improvement must be made in accordance with the scientific method, under the guidance of a teacher, at the lowest possible level in the organization." (Spear and Bowen 1999)

TPS centers the production system around the person conducting the work and giving him/her more autonomy (rule 4, "at the lowest possible level"). The scientific method (rule 4) demands short, self-organizing feedback loops which involve the worker in that he/she assesses his/her workspace and provides ideas for improvement. In order to see changes made to existing processes, TPS strictly applies the principle of standard work (rule 1). Standard work is defined with participation of the worker and standard work instructions can be revised. The current version of standard work must be followed. Definition of standard work enables transparency of deviations from the planned production process. Identified deviations trigger investigation. Improvement efforts are rooted in the principle of experimentation: revisions of standard work can be seen as an experiment to the status quo which may lead to an improvement. Changes to the status quo must enable direct relations between customer and supplier (rule 2) and enable simple and direct pathways (rule 3). An open, no-blame culture supports the principle of experimentation and regards breakdowns as a chance to learn. Experimentation supports challenging the status quo of the production system. The principle of investigation supports experimentation by providing tools for problem identification and solution finding. These tools provide structured processes for thorough and collaborative investigation.

To summarize, while the TPS is very rigid regarding the rules for conducting improvement, i.e., how change happens, the actual work processes, i.e., what changes, become dynamic over time, also because of decentralized control over improvement efforts. TPS supports the worker in improving by giving him the means to do so, i.e., TPS manages "by means" (Rother 2009, p. IX).

2.3.2.2 Lean Construction

The concept of TPS has been transferred into a broader industry context (Womack and Jones 2010) and applied outside the automotive industry. One such transfer was the adaptation of lean production to the construction industry. Lean construction uses the same concepts and principles as lean production, but adds methodologies which are geared towards the characteristics of project production.

The Lean Construction Institute (LCI) developed the LCI triangle (figure 26), which structures the construction production system into three domains: organization, commercial terms, and operating system (Ballard 2012; Howell et al. 2011; Thomsen et al. 2010). These three domains can affect the use of technology in a project; here technology includes methods and tools, for example BIM.

The Operating System domain includes planning and control of work. The Organization domain includes assignment of responsibilities for process steps, establishing the organizational structure through protocols for vertical and horizontal communication, project culture, and leadership style. Here, the organization domain influences the operating system domain, e.g., by enabling continuous improvement efforts through flexible organizational structures and a collaborative project culture. The assignment of responsibilities for process steps influences the structure of commercial terms of a project; in turn the setup of commercial terms sets incentives for optimizing parts of the project vs. optimizing the project as a whole. The Commercial Terms domain includes the contracts established between all parties involved in a project as well as rules from other sources in the environment of the project.

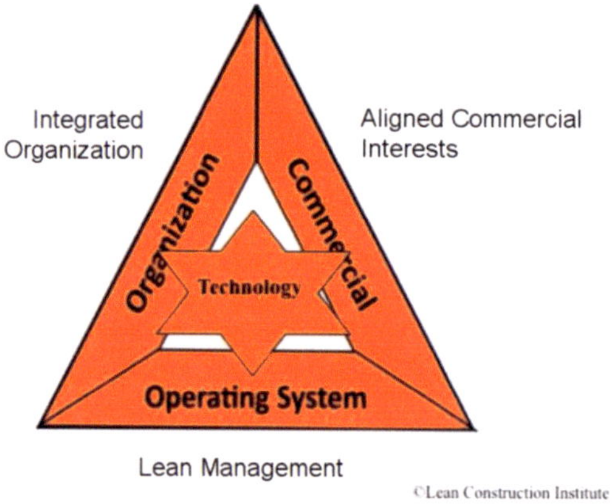

Figure 26: LCI Triangle (Ballard 2012)

Howell et al. (2011) compare two approaches based on the three domains (table 11). In traditional Project Delivery the Operating System is activity-based, i.e., management applies the logic of the transformation perspective of the TFV theory. In Lean Project Delivery the Operating System is flow-based, i.e., management applies the logic of the flow perspective of the TFV theory. In Traditional Project Delivery the Organization is often hierarchically structured into silos and it follows an authoritarian command and control management style. Lean Project Delivery integrates the organization by avoiding silos and it installs a collaborative management style. In Traditional Project Delivery the Commercial Terms mostly follow the transactional approach, in which two parties agree on a transaction of an object for money with characteristics of the object fully specified in advance. Relational contracts define the relation between the parties. They can be set-up as multi-party contracts between more than two parties. The purpose is to align the commercial interest of the parties involved (see section 2.3.2.2 for more details). The approaches described here are not exhaustive; instead they can be regarded as two points on a continuous scale: several mixed approaches exist. See Lahdenpera (2012) for a comparison of three approaches for the domain Commercial Terms.

Table 11: Domains of Project Delivery (Howell et al. 2011)

	Operating System	Commercial Terms	Organization
Traditional Project Delivery	Activity Centered - CPM	Transactional	Command and Control
Lean Project Delivery	Flow – Lean based	Relational	Collaborative

Thomsen et al. (2010) state that the characteristics of the three domains must be aligned for successful project execution. Imbalanced approaches to project delivery systems are less successful than balanced approaches, when considering the project as a whole. This observation highlights interdependencies between the three domains.

Lean construction adds at least three methodologies to the existing lean production methodologies: (1) Lean Project Delivery System, (2) the Last

Planner System™, and (3) Integrated Project Delivery as a contractual means for setting a project up in a collaborative manner.

(1) Lean Project Delivery System (LPDS)

LPDS (figure 27) is a procedural model of the lifecycle of a building which is structured into five overlapping phases, represented by triads: project definition, lean design, lean supply, lean assembly, and use. Throughout all phases the LPDS applies production control, e.g., with the Last Planner System™, and work structuring (Ballard et al. 2001a; Tsao et al. 2000). The LPDS is based on three principles regarding work structuring (Tsao et al. 2000):

- Integrated product and process design: the 'lean design' triangle contains the tasks 'design concepts', 'product design', and 'process design'. The principle of integrated product and process design is rooted in the concept of concurrent engineering (CE), which has also been transferred to the AEC industry (Anumba and Evbuomwan 1997; Gunasekaran and Love 1998). CE aims at integrating all relevant criteria for decisions regarding product and process design, instead of executing these phases sequentially. Hence, from a building lifecycle perspective, CE integrates downstream knowledge early.
- Work structuring together and early: collaborative and joint programming increases the quality of plans and reduce the probability of process breakdowns. Joint programming integrates parallel processes in order to find dependencies in advance and to establish a continuous workflow.
- Continuous improvement as an integral part of all processes: learning loops occur not only between projects but in short feedback-cycles within each triad.

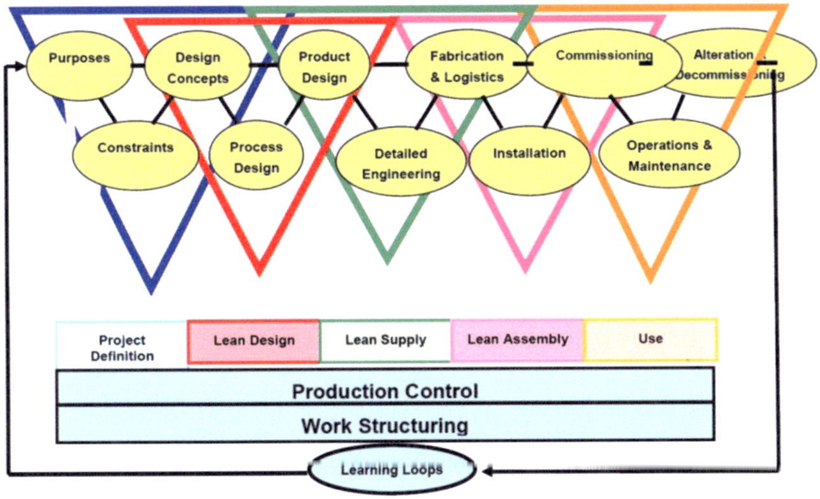

Figure 27: Lean Project Delivery System (Ballard 2000a, 2008)

(2) Last Planner System™

Ballard (1994; 2000c) describes the LPS as a means for production control. The name derives from a goal of the LPS: to involve the Last Planner in production control. The Last Planner in construction is usually the foreman who plans detailed work processes on-site.

The LPS consists of four phases and is based on one additional phase, the master schedule. The purpose of the first three phases is to enable collaborative planning with a gradual increase in planning detail, i.e., the closer work comes to its execution the more detailed it is planned.

(1) The phase Schedule specifies hand-offs between work-packages. The team develops the phase schedule between two milestones which stem from the master schedule.

(2) The look-ahead-Schedule defines tasks, assigns responsibilities, and makes tasks ready for execution by removing constraints.

(3) The weekly Work Plan releases constraint-free tasks for execution.

(4) Learning measures the reliability of production by calculating the PPC value. It compares tasks executed to tasks released. Transparency regarding actual task completion fosters learning that acts on failures through root-cause analysis and investigation.

Throughout all phases Last Planners drive the process and make commitments to each other. Last Planners have the possibility to deny an assignment, they can say "no". Making commitments follows Flores' coordination cycle (Flores 1981, p.78); commitments establish direct customer – supplier connections and they specify the characteristics of the task in accordance with both customer and supplier.

The Last Planner System installs a participatory project leadership style that adds the planning states "can" and "will" to production management (Ballard 2000c, p. 3-2) (figure 28). Responsibility for work structuring moves partially to the people, who execute work, the Last Planners (Ballard 2000c, p. 3-14). The LPS establishes a mix of bottom-up and top-down management: the master schedule sets top-down constraints for production while Last Planners plan operations bottom-up within these constraints.

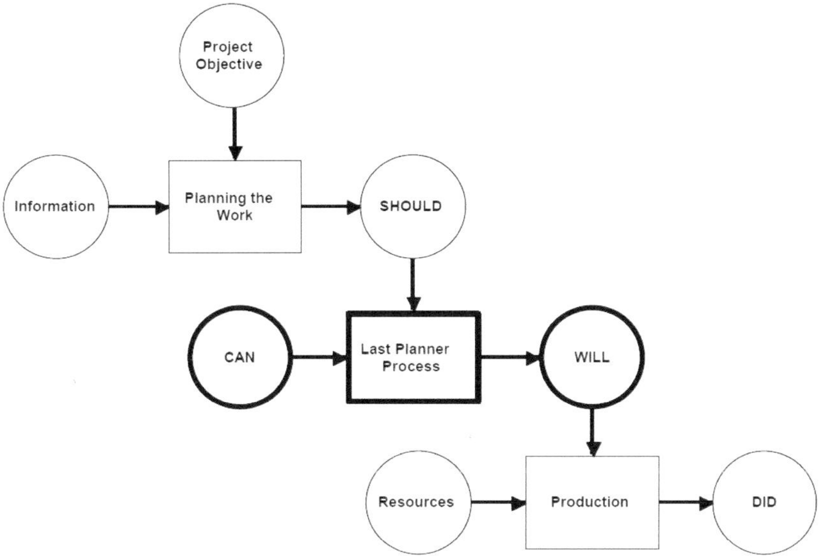

Figure 28: The Last Planner System (Ballard 2000c, pp. 3-15)

The LPS has been successfully applied to a large number of AEC projects. Cho and Ballard (2011) show the positive correlation between an extended use of the LPS and cost and schedule reductions. Mossman (2015) presents anecdotal evidence of 30% productivity improvement through LPS application. The LPS has also been successfully applied to the design phase of AEC projects (Ballard 2002; Hamzeh et al. 2009).

(3) Integrated Project Delivery[17]

Projects are temporary socio-technical systems, completed usually not by an individual, but by a group of people who must interact. This interaction is influenced by the characteristics of the project delivery system (Thomsen et al. 2010).

[17] Part of this section has been published in Hickethier et al. (2013).

Commercial Terms and specifically the relational contract terms of IPD projects promote collaboration between project members by including mechanisms such as pain-and-gain sharing, collective risk management, and contingency sharing. These mechanisms affect the relations between project members and promote strong collaboration (Howell et al. 2011; Thomsen et al. 2010).

The Operating System of IPD projects is based on the principle of reliable workflow (Howell et al. 2011). Key practices for increasing the reliability of information flow in design use, e.g., learning through PDCA thinking and root-cause analysis, look ahead planning with the Last Planner System™, Value Stream Mapping (Rother and Shook 2003), and Target Value Design (Zimina et al. 2012).

Project organizations that follow an IPD agreement integrate owners, designers, and contractors. Contractors join the design team early and all partners work from a collocated office. Integration of knowledge across trades and disciplines and across the building lifecycle enables opportunities for increased value generation.

Cross-functional teams consisting of individuals from the relevant companies find innovative and efficient solutions through their diverse set-up. An executive committee consisting of members from the involved companies manages the teams, makes decisions unanimously through consensus, and creates an open, collaborative culture. This model creates a 'virtual company' (Thomsen et al. 2010) with members employed by their home companies but trusting each other. The resulting collaboration fosters the behavior that the best qualified person does a job, regardless of their home company. Table 12 summarizes the structural characteristics of IPD design organizations. Early involvement of contractors and integrated organization both focus on the set-up of the overall organization, while flexibility presents a dynamic capability of the organization.

Table 12: Structural Characteristics of IPD Design Organizations

Structural Characteristics of IPD Organizations	**Description and References**
Early involvement of contractors during the design phase	Contractors, designers, and owners are involved from the early stages of the project (Thomsen et al. 2010, p.11).
Integrated organization	Contractors, designers, and owners interact during design (Thomsen et al. 2010, p.11).
Flexible organization	In IPD projects people are encouraged to do what is best for the project (Heidemann and Gehbauer 2010), they become part of a virtual company (Thomsen et al. 2010, p.11). The mix of top-down and bottom-up management with LPS encourages people to promote improvement of the production system (Gehbauer 2008). The team uses standardized but flexible procedures which are subject to improvement (Thomsen et al. 2010, p.44). Global optimization of a project demands that distribution of project scope is flexible and money related to that scope must be able to move across contractual and organizational boundaries (Ballard 2012).

2.3.2.3 Lean Design

Lean Design is the application of Lean Management to design processes. Principles and methodologies have been developed specifically for Lean

Design. Uncertainty is an inherent part of the design process, and it surfaces also in iteration. While iteration in production is wasteful because it represents rework, iteration in design can be value-adding. Parameters that describe a building or product are often interdependent; sometimes reciprocal dependencies between several parameters exist.

Lean Management focuses on the Flow and value perspective on production[18] (Ballard and Koskela 1998; Koskela 2000). Next follows a more detailed description of the flow and value perspectives in Lean Design and related organizational principals which support Lean Design.

(1) Lean Design from the Flow Perspective

Research in Lean Design Management has developed methodologies for the flow perspective (Koskela 2000), e.g., Theory of Constraints (Goldratt 1990), Toyota Product Development System (Morgan and Liker 2006), Lean Product Development Flow (Oppenheim 2004), Product Development Value Stream Mapping (McManus 2005), the Design Structure Matrix (Tuholski and Tommelein 2008), and Lean Design in Lean Construction (Freire and Alarcón 2002).

Management of information flow focuses on the reduction of waste; however, a difference between waste in production and waste in design exists. In production, Ohno (1988, p.19) differentiates between value adding, necessary, and wasteful tasks. Value adding tasks contribute to the delivery of customer value, while necessary tasks and wasteful tasks make no value contribution. Value delivery would be impaired without completion of necessary tasks. But wasteful tasks can be removed from the process without impairing value delivery. Thus, value adding tasks shall be optimized, necessary tasks shall be minimized, and wasteful tasks shall be removed.

[18] See section 2.3.1 for a description of the TFV theory.

However, the transfer of this concept to design and development processes demands considerations regarding the nature of the design process: product-related uncertainty in design hinders a priori differentiation between value adding and necessary tasks, because the value concontribution of design tasks can often only be evaluated in retrospect (Browning 2003). Nevertheless, Ohno's (1988, pp.19f.) seven kinds of waste, a tool for analyzing and improving processes, can be transferred to design. Koskela (2004) adds 'making do' as an eighth kind of waste, which refers to the initiation of an activity without all necessary inputs available. Macomber and Howell (2004) add 'not speaking' and 'not listening' as the ninth and tenth kind of waste. Table 13 gives examples from AEC design for each of these ten kinds of waste. The additional kinds of waste enhance Ohno's (1988, pp.19f.) original classification of waste.

Table 13: Classification of Waste and AEC Examples (based on Tuholski (2008, p.46) and Macomber and Howell (2004))

Waste Classification	AEC Example
Overproduction	Completing a design package early, before it is needed in the field or shop.
Waiting	Steel beam design awaiting piping layout.
Transportation	Shipping project drawings.
Processing Itself	Hand-marked sheets that are thrown out after drafting.
Inventory	Backlog of red-marks awaiting drafting.
Movement	Emailing design parameters.
Defective Products	Design errors due to mistake or improper application of criteria.
Make Do	Designing an element out of sequence because the inputs to the properly sequenced work were not available or assumptions were in error.
Not speaking	Not raising an important issue, because a person's experience tells him/her that criticism is not well received in this project.
Not listening	Designers suggest process improvements but the project manager does not listen.

Several authors have provided frameworks for the classification of waste in mechanical engineering design (e.g., Bauch 2004; Morgan and Liker 2006; Pessôa et al. 2009; Shah and Ward 2007). Bauch (2004) and Pessôa et al. (2009) differentiate different kinds of waste into generic sources of waste and analyze which sources of waste trigger other sources. The resulting network of sources of waste reveals that rework of activities is often caused by other sources: rework is often an effect of other sources of waste. They analyze the cause and effect chains with DSM in order to determine strategies for waste removal.

Browning (2003) identifies the risk of removing non-wasteful activities during waste reduction, when the focus of waste reduction is on the individual activity and not on the overall process. Especially when uncertainty hinders the differentiation between necessary and wasteful activities, the minimization of necessary activities is potentially counter-productive as it may reduce value generation. Browning (2003) focuses on the structure of the overall design process, instead of focusing on specific activities: "[..], the architecture of the PD process – the sequencing and coordination of activities and their deliverables – has a large impact on value, regardless of the value of the activities and deliverables themselves." He recommends focusing on the improvement of value generation in NPD through a better structuring of the process and a subsequent effective coordination during execution of tasks rather than the removal waste. Ballard (2002) shows that the Last Planner System™[19] is an effective tool for coordination of task execution in design projects.

Ballard (2000) addresses the iterative nature of processes in design. He distinguishes between positive and negative iteration. He defines negative iteration as waste "which can be eliminated without loss of value or causing failure to complete the project." Tribelsky and Sacks (2010) show that iteration can cause rework, thus increasing project cost and duration. Ballard (2000b) provides 12 strategies for the reduction of negative iteration.

The mode of information transfer between tasks impacts the design process flow: suppliers can push information towards customers, or customers can pull information from suppliers. Morgan and Liker (2006, p. 96) describe the pull mode of information transfer in the Toyota Product Development System:

"in product development, knowledge and information are the materials that are required by the downstream activity. However, not all

[19] Section 2.3.2.2 describes the Last Planner System™.

information is equal to all people. The lean PD System uses 'pull' to sort through this mass of data to get the right information to the right engineer at the right time. Knowledge is the fundamental element (material) in product development. Toyota does very little "information broadcasting" to the masses. Instead, it is up to the individual engineer to know what he or she is responsible for, to pull what is needed, and to know where to get it."

(2) Lean Design from the Value perspective

From the value perspective Lean Design has developed several methodologies. Table 14 describes the methodologies set-based design, rundown of requirements, and Target Value Design.

Table 14: Description of Lean Design Methodologies from the Value Perspective

Methodology	**Description**	**References**
Set-based design	The design team researches the whole set of design alternatives and gradually narrows the set based on customer value.	(Hickethier et al. 2011a; Parrish 2009; Ward et al. 1995)
Rundown of requirements	The design team focuses on the voice of the customers, e.g., end users and production, and analyzes and prioritizes their requirements in a structured manner. The team deduces requirements beginning with the end-user through several layers until production planning with Quality Function Deployment (QFD).	(Cristiano et al. 2001; Delgado-Hernandez et al. 2007)
Target Value Design	The design team defines target values for a design project, including cost, time and capabilities of the product. These values are the foundation for decision making during design, instead of being the result of design.	(Ballard 2011; Ballard and Reiser 2004; Zimina et al. 2012)

(3) Organizational Roles and Structures in Lean Design

A large number of different approaches to roles and organizations exist in Lean Design. Table 15 describes roles and structural characteristics that are often applied during the design phase of projects that apply Lean Construction.

Table15: Organizational Roles and Structures

Role or structural characteristic	**Description**	**Reference**
Chief Engineer (CE)	The CE has high informal power within the organization through respect and experience, but he/she has little formal authority. However, he/she is responsible for the results of the project and he/she leads the team by focusing efforts on delivery of customer value.	Morgan and Liker (2006, p.132)
Cross-Functional Team	Cross-Functional teams consist of specialists who work in a matrix organization under a balanced leadership of functional organization and product organization and who focus their work on delivery of customer value.	Morgan and Liker (2006, p.145ff.)
Collocation	Collocation focuses on organizational integration by locating workplaces of people involved close to each other, e.g., in one big office.	Kahn and McDonough (1997)
Big-Room (Obeya)	A designated meeting room with visualizations of all important project information including schedules and key metrics.	Morgan and Liker (2006, p.152 f.)

2.4 Summary

Chapter 2 covered a number of fields of literature and revealed several domains of complexity that impact the AEC design system. A production system perspective on AEC design must include these impacts. This perspective provides at least three characteristics of AEC design systems:

(1) Constructivism impacts performance of the AEC design process, especially when people change between projects and the constructed realities differ due to different professional backgrounds.
(2) Product-related uncertainty causes ongoing definition and concretization of requirements for the building throughout the design process.
(3) Projects are open systems which interact with their environment.

TPS and Lean Construction are based on the scientific method. The scientific method fosters learning by finding root-causes for deviations from planned outcomes. This type of reflection leads to self-reference of the system. When the scientific method is applied, it promotes self-organization and thereby facilitates system emergence.

The set-up of the AEC design system influences the structure of the communication network between people. The next chapter synthesizes the research gap by detailing the research focus and by analyzing methodologies for reflection of communication structures.

3 Research Gap

Building on the literature presented in Chapter 2, this chapter presents the research gap and it is structured as follows: Section 3.1 summarizes the research focus for identifying the research gap. Section 3.2 analyzes the research gap. Section 3.3 presents requirements for filling the research gap, and section 3.4 summarizes this chapter.

3.1 Research Focus

Communication is the focal point of research in this dissertation, and communication must be analyzed in the context of project and environment. The organization, in which people communicate, is set within a larger project context, which again is set within a project environment. Models of design projects and project environment were presented in chapter 2. These kinds of models of the design system serve the purpose of enabling users to better understand the interdependencies within the system and these models stem mostly from mechanical engineering design. These models are often generic in nature, and thus translatable to AEC design. Figure 29 fuses three models in order to combine the following characteristics of these models in the context of AEC design:

- construction projects are open systems. A project interacts with an environment, culturally, and otherwise, and preceding events in time affect the project (also called path dependence) (Engwall 2003).
- a project can be partitioned into four domains: (1) product, (2) commercial terms, (3) organization, and (4) operating system. The three domains of organization, commercial terms, and operating system shall be aligned regarding the management approach (Bal-

lard 2012), which shall be aligned with characteristics of product and environment.

- design is an ongoing concurrent development of ends and means. That is, the organization develops - through its operating system within existing commercial terms - product representations (object system), which help owners, users, and other customers of the design process in learning about product and project requirements (system of objectives), thereby reducing product-related uncertainty (Albers and Meboldt 2007).

Communication takes place within the project system, and this system is subject to uncertainty. Communication is a means for coordination, but also a prerequisite for improvement of a socio-technical system (Baecker 2003, p.21; 2006). As project systems are subject to uncertainty, the project must re-organize constantly to adapt to a changed situation (Bahrami and Evans 2011). Furthermore, projects must re-organize constantly to improve themselves (Baecker 2003, p.19). Then, projects are no longer static, but "becoming" (Koskela and Kagioglou 2006b) and management then includes constant observation and re-drawing of the boundary between project and environment (Baecker 2003, p.227f.). Therefore, the boundary between project and environment is not only permeable but also fluent, as indicated by the dashed line in figure 29.

Improvement and adaptation demand re-organization. The project develops new patterns of communication (Gehbauer 2008), which necessitates flexibility within the project organization. Flexibility can be defined as the ability "to move rapidly, change course to take advantage of an opportunity or to sidestep a threat" (Bahrami and Evans 2011).

Different types of project organizations exist, and the type of commercial terms (including contract) impacts the structure of the project organization by defining, e.g., lines of reporting between project participants. Projects that use an IPD-type contract have been associated with flexible organization structures (table 12). This dissertation aims at (1) illuminating the organization structure of IPD projects, and (2) providing a method for improving communication structures. The following sections will provide a deeper look into communication and uncertainty in organizations as well as work structuring as a method for developing communication structures in design.

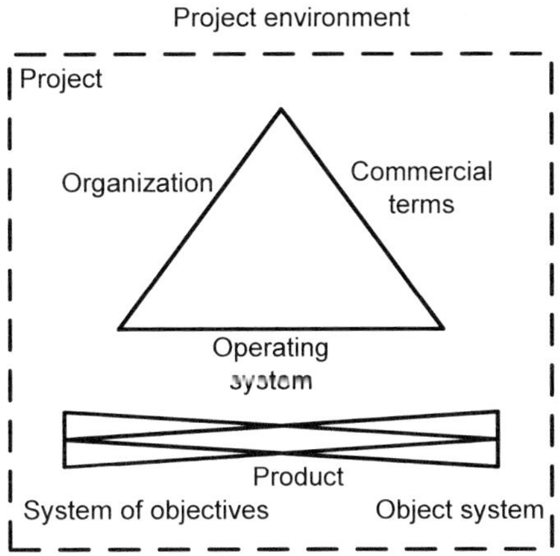

Figure 29: System Model of the AEC Design Process (based on Albers and Meboldt (2007); Ballard (2012); Engwall (2003))

3.1.1 Communication Structures in the AEC Design System

People coordinate their work through communication, but communication is also a necessary vehicle for learning and improvement. Baecker (2003, p.62) argues that communication is the core capability of the Toyota Production System (TPS) for continuous improvement. Visualization serves as a vehicle for communication, and this communication has the purpose of enabling quick reaction in case of disturbances in the production line. Visualization of production results enhances self-reference[20] of the organization by achieving transparency. Through visualization people communicate what they are doing, when they are doing it, and what the results are. Also, people can more easily observe the outcome of their actions. The ability to observe results and disturbances enhances the ability to reflect own actions and to analyze reasons for disturbances. Reflection and analysis spur learning for continuous improvement of production by reducing waste. Reduction of waste, e.g., through reduction of buffers, makes the production system more fragile, but this fragility increases its robustness, because it improves the ability to react quickly in case of disturbances (Baecker 2003, p.63).

Communication and autonomy to change behavior are enablers for continuous improvement (Baecker 2003, p.27). Lean Management implements communication and autonomy through a set of rules. Spear and Bowen (1999) describe four rules of the TPS (see section 2.3.2.1). Rother (2009, p.176) describes the goal of autonomy in TPS as to "empower or engage process operators" in improvement efforts, but not to allow self-directed teams. At the core of improvement lies the scientific method, which demands short, self-organizing feedback loops which involve the worker in that he/she assesses his/her workspace and provides ideas for improvement.

The TPS as well as the LPS propose deutero learning, i.e., learning from failures through observation, reflection, and analysis of results. Learning means developing new structures of communication and also discarding established routines (Gehbauer 2008).

Flexibility is a critical capacity for developing new structures of communication. Bahrami and Evans (2011) identify clear boundaries and autonomy of workers as a prerequisite for flexible organizations. Furthermore, they identify three domains of flexibility: reporting relations,

[20] Self-reference is a principle of General Systems Theory; Baecker (2003, p.226) argues that his social management theory is rooted in 2nd order cybernetics (see section 2.1.3.1).

organization culture and identity, and lateral relations. Bahrami and Evans (2011) name lateral relations as the "critical lever [...] to orchestrate rapid changes that can cumulatively reshape an entity over time". This statement is in line with Gehbauer (2008), who proposes to use the LPS in projects to facilitate gradual changes in the line organization that operates the projects.

3.1.2 Uncertainty in the AEC Design System

Baecker (2003, p.36) describes the need to communicate uncertainty in order to avoid accumulation of risk. Worren (2012, p.201) describes how levels of product-related and process-related uncertainty (see section 2.1.2) in a project impact the degree of work level interdependencies. However, both types of uncertainty are not a given for a project, but their levels are within control of the project. Process reliability influences process-related uncertainty, and the LPS can increase process reliability (Ballard 2000c). Transparency of customer requirements relates to product-related uncertainty, and methods, e.g., TVD (Zimina et al. 2012), can increase transparency and accelerate the process of reducing product-related uncertainty. At the end of the design process, which is defined by the existence of a product model (e.g., building plans), and a recipe for producing it (e.g., construction sequences), product-related uncertainty is sufficiently low to start procurement and construction.[21]

Figure 30 presents a model that relates the degree of work level interdependencies with the formal organization structure. The degree of work level interdependencies depends on uncertainty and importance of information. Both impact the characteristics of process management. A team can define processes far in advance, when uncertainty is low. But high uncertainty can inhibit a team's ability to foresee dependencies

[21] In practice, end of design and beginning of production often overlap. End of design is often not clearly defineable, especially when late design changes occur after the design was assumed to be finished.

between tasks, thereby hindering process definition. Customer-supplier connections between tasks are less predictable when uncertainty is high. In this case, efficient process execution requires integrated and flexible organization structures. Integration increases information exchange by increasing the range of recipients for information. This provides people the opportunity to receive information that allows them to identify dependencies, thereby reducing uncertainty. Flexibility supports adaptation of the organization so it can conduct a process that accommodates the newly identified dependencies.

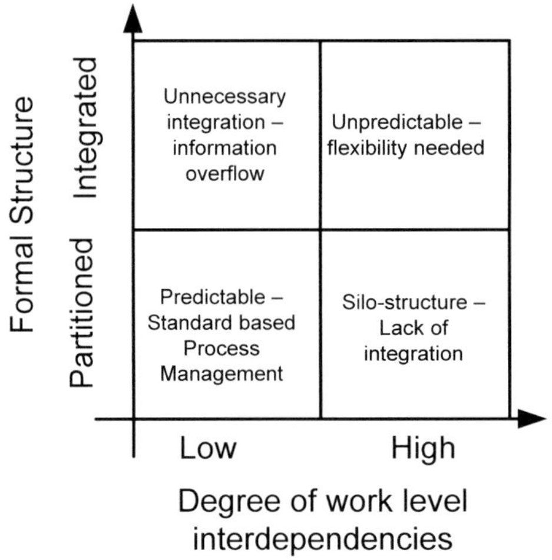

Figure 30: Relationship between Uncertainty and Integration; based on Worren (2012)

The goal of design processes is to define a 'recipe' for the product, i.e., to eliminate product-related uncertainty regarding requirements and solution, or ends and means, by the end of the design process. As product-related uncertainty decreases, product specificity increases. Hence, it can be assumed that the level of product-related uncertainty is not static throughout the design process, but dynamic instead. Since product-related uncertainty influences the formal structure, adaptation of the design system is an integral part of design. Adaptation is usually associated with change and organization contingency theory presents a number of different theories. Rother (2009, p.168) describes that the ability of a firm to survive is related to its ability to adapt and change. The goal of adaptation is related to the concept of improvement as described in the value perspective of the TFV-theory, which is delivery of customer satisfaction. Hence, organizational adaptation can be subsumed under the general concept of improvement as described in the TFV-theory[22].

It can be assumed that the design process reduces levels of product-related uncertainty over the run-time of the design phases. At the end of the design process, product specificity has reduced product-related uncertainty sufficiently to begin construction. The dynamic levels of uncertainty affect the need for organization integration (figure 31) and lead to an emergent partitioning of the organization. Adaptation can also be necessary due to changes in the project environment, improvement of structures, or other reasons.

[22] See section 2.3.1 for a description of the TFV theory.

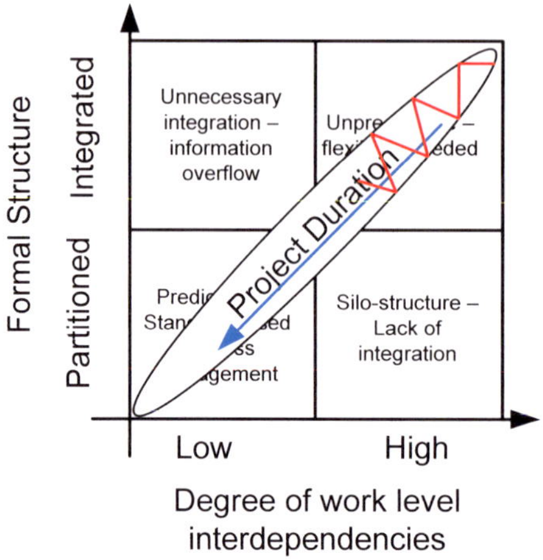

Figure 31: The Impact of Reduction of Product-related Uncertainty over Project Runtime; partially based on Worren (2012)

As the need for integration changes over the run-time of a design project, so does the organization architecture.

3.1.3 Work Structuring in Design Organizations

Work structuring (Ballard 1999; Tsao et al. 2004) (see section 2.2.2.3) partitions the project scope into smaller units and then reintegrates these units by defining work sequence, work release, workflow, buffering of workstations, and production schedule. Thereby, work structuring is a part of production system design. Work structuring installs a set of rules for production, and these rules aim at reducing process-based uncertainty. Work structuring partitions and integrates project scope from a process perspective with a focus on lateral relations.

Browning (2009) presents Design for Integration (DFI) (see section 2.2.3.2) which consists of six steps: understand product and process system structures, assign integrated product teams to product or process entities, group integrated product teams, apply integrative mechanisms, manage interfaces, and reassess status. DFI designs lateral relations with a focus on organization architecture, and integrative mechanisms.

Work structuring (Tsao et al. 2004) and Design for Integration (DFI) (Browning 2009) are compatible and can complement each other. Work structuring focuses on the process, while DFI focuses on organization architecture and integrative mechanisms. Process, organization architecture, and integrative mechanisms establish lateral relations within an organization.

The focus of this dissertation lies on communication in the project organization. Organization partitioning, integrative mechanisms and process execution influence lateral relations between entities of the organization. Work structuring establishes a process model which shows, among other attributes, a sequence of work. The sequence of work prescribes a pattern of communication. Integrative mechanisms and partitioning of the organization architecture similarly prescribe patterns of communication. Communication prescribed by process, organization architecture, and integrative mechanisms is a subset of the overall interaction network of an organization. Relationships of authority, responsibility, accountability, and others also influence to the overall interaction network.

Figure 32 combines DFI (Browning 2009) work structuring (Tsao et al. 2004) and sets them in the context of the PDCA cycle (Deming 2000, p.88) to develop a model for planning and improvement of communication structures. This model focuses on lateral relations established by process, organization architecture, and integrative mechanisms. The model also builds upon the PDCA cycle (Deming 2000), which partitions the model into the four parts 'Plan, 'Do', 'Check', and 'Act'. The 'Plan' section integrates planning of process (how will work be sequenced?), organization architecture, and integrative mechanisms. The 'Do' section of the model focuses on executing planned work, and the 'Check' section focuses on evaluation whether prescribed communication patterns were followed during execution of work. The 'Act' section of the model focuses on changing characteristics of each step of the model.

The first five steps plan the communication structure of the design organization. Steps one to four focus on integration (steps from (Browning 2009)) and step five focuses on coordination (from (Tsao et al. 2004)). Step six makes use of the communication structures by 'do'-ing work. Step seven 'check's the status of communication structures in the context of the conducted work. Following step seven, people 'act' by re-planning communication structures. 'Act' establishes a feedback loop of communication structures, and the method presented in this dissertation aims at improving this feedback. The method described in this dissertation relates to step seven "evaluate status".

The model presented in figure 32 focuses on the structural attributes of communication-type relations between people. The model disregards some attributes of communication-type relations, such as mode of information transfer (push/pull) and batch size of information transfer. These attributes are important for managing efficient execution of the process. This research elaborates a method for comparing process structures. Mode and batch size of information transfer are not within the scope of the method elaborated in this research.

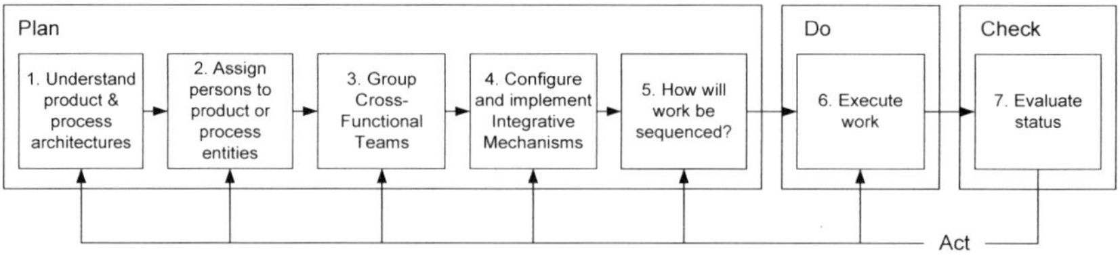

Figure 32: Model for Planning and Improvement of Communication Structures

More specifically, the focus of this dissertation lies on the 'check' part of the PDCA cycle, evaluation of status of the communication structures.

The following sections describe the research gap regarding the evaluation of communication structures in detail.

3.1.4 Methods for Planning and Improvement of Communication Structures

Management consists of planning, execution, and controlling of processes (Koskela 2000, p.27). The Lean Construction approach to project management focuses on the improvement aspect of controlling. The method elaborated in this research aims at improvement. Therefore, the following sections focus on the improvement part of controlling.

Planning and improvement of processes relate to the definition and redefinition of communication structures; thus they are the focus of this section. Execution is tightly connected to planning and controlling, and thus deserves a short review.

Proponents of Lean Construction advocate the LPS (Ballard 2000c) as the method of choice for process execution also in design (Ballard 2002; Hamzeh et al. 2009). The LPS also has planning and improvement characteristics that will be reviewed later. The LPS (Ballard 1994, 2000c) advocates process coordination based on Flores' (1981) LAP[23], and measuring process reliability. LAP serves as the model and definition of communication in this dissertation.

Several methods for planning and improvement of communication structures have been described in literature. Figure 33 presents an overview of existing methods for domain spanning analysis based on structural modeling. The review of existing methods focuses on domains product, process, and organization. The commercial terms domain, as the fourth domain of the systems model, is excluded from this review. The

[23] Ballard's (2000c) original description of the LPS does not mention LAP, but it was later added by a series of papers (Howell et al. 2004; Macomber et al. 2005; Macomber and Howell 2003).

structure of commercial terms is set-up at the beginning of the project and it serves as the starting point for structuring the organization domain. Contractual obligations connect companies. It is assumed that the commercial terms are sufficiently represented in the organization domain, and therefore the domain 'commercial terms' is excluded from this review.

References placed in the center of the triangle focus on all three domains; references placed on the edges focus on the two respective corners of the triangle. The following sections structure the presented literature regarding their focal domains and purpose.

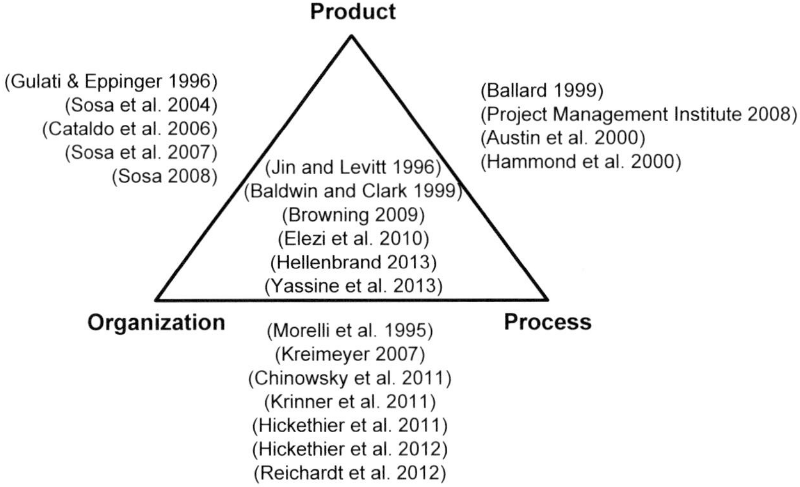

Figure 33: Methods for Domain Spanning Analysis of Structures

3.1.4.1 Planning of Communication Structures

Based on the scope of research of this dissertation, planning of communication structures includes:

- planning of the process structure,
- planning of the organization architecture,
- planning of integrative mechanisms.

Planning of communication structures often takes the product structure as a starting point. Jin and Levitt (1996) develop a simulation based on coordination requirements (derived from product complexity and uncertainty) and coordination capacity (based on process architecture, organization architecture, and integrative mechanisms) to improve project planning. Baldwin and Clark (2000, p.48) explain that the structure of design tasks shall mirror the structure of dependencies of design parameters of the product. Further, they explain that organization structure shall mirror task structure (Baldwin and Clark 2000, p.54). Browning (2009) extends the approach of Baldwin and Clark (2000) with directions for implementation and by including 15 integrative mechanisms. Elezi et al. (2010) extend Baldwin and Clark's (2000) approach by improving process and organization architecture separately with algorithms before deducing them into the next domain (based on Maurer (2007, pp.82f.)). Hellenbrand (2013) extends Elezi et al.'s (2010) approach with tools for the identification of change effects and product maturity. Yassine et al. (2012) present an algorithm for the global optimization of all three structures.

Gulati and Eppinger (1996) propose a mirroring between product architecture and organization architecture; they discuss the effects of several integrative mechanisms on product characteristics and communication.

The Project Management Institute (2008) derives the process structure from the product architecture: WBS partitions the product and then assigns tasks that develop or build these chunks while still giving regard to dependencies between the tasks. Ballard (1999) presents the concept of "work structuring" which extends WBS towards production system design. Austin et al. (2000) present the Analytical Design Planning Technique (ADePT), which combines task-based DSM with WBS. Hammond et al. (2000) extend ADePT with a software instantiation of the

LPS. Following the idea of the process-based organization, Morelli et al. (1995) predict communication from a process structure. Krinner et al. (2011) deduce organization architecture from process structure by using a LPS phase plan as a model of the process structure.

To summarize, the presented methods focus on deduction of structures from existing structures. The starting point is often a modular product architecture, which is deduced either directly into the organization domain or through the process domain into the organization domain. The presented methods focus mostly on planning communication structures through deduction of structures, and improving communication through better planning. The next section focuses on methods for improvement of communication through comparison of structures.

3.1.4.2 Improvement of Communication Structures

Communication transports information which serves as input for tasks. Tasks transform inputs into outputs and the goal of transformation is to increase information value. Communication as the vehicle for information flow is also subject to the TFV-theory of production which provides three different perspectives on improvement (table 16).

Table 16: TFV Perspective on Improvement

Perspective on Production	Focus of Improvement
Transformation	Increased productivity through innovation (Koskela 2000, p.45).
Flow	Elimination of variability (Koskela 2000, p.64).
Value	Customer satisfaction (Koskela 2000, p.82).

From a systems perspective, learning and improvement demand self-reference of the design system. Self-reference enables a system to emerge. Several scholars from different fields of research highlight the importance of self-reference and its configuration for emergence, e.g., project management theory (Koskela and Howell 2002), theory of complex adaptive systems (Dooley 1997), and organization theory (Baecker 2003, p.21; Pulm 2004, p.123).

Emergence of structures necessitates self-reference[24] within the system, i.e., the system needs feedback regarding its own state. Pall (2000, p.165) describes this pre-requisite of process improvement:

> "In all business processes, a transition towards a new state is only possible if information has been imported into the process. This is the concept of feedback, and it means that the process can maintain and move itself to higher levels of capability only with the aid of feedback information. The prerequisite for continued process capability change is the availability of adequate and timely feedback information representing changes occurring in the environment of the process."

Different types of feedback exist; Koskela and Howell (2002) describe two typical approaches for feedback:

(1) Cybernetic model of management control or "thermostat model": the model compares planned performance to measured (actual) performance. If there is a variance between the two values, actions are taken to correct the process and set it back on track, so that planned performance can be achieved.

(2) Scientific experiment model: the model specifies operations, poses hypotheses, runs the operation, and then tests the hypothesis by comparing it to results. Specification enables root-cause analysis, which is conducted, if hypothesis and result deviate. This model applies the Lean Management principles of 'experimentation' and

[24] Section 2.1.4.1 contains a description of self-reference as a principle of General Systems Theory.

'investigation' (section 2.3.2.1) and is an integral part of the TPS (Spear and Bowen 1999).

A review of current literature identified three different classes of methods for achieving feedback. Feedback usually paints a picture of the current state or past of a system, and methods for feedback usually present only a subset of attributes of that system. The three classes of methods for feedback differ, based on their focus (i.e., what attributes of the system they present) and how they gather information.

(1) Feedback regarding performance, e.g., time, cost, and delivery of customer requirements. These metrics are also described in project management literature (e.g.,(Project Management Institute 2008)). O'Donnell and Duffy (2005, pp.195ff.) present a comprehensive overview of metrics for measuring design performance. Bashir and Thomson (1999) review metrics for calculating development time and cost, which can serve as a basis for comparison with actual values.

(2) Feedback regarding actual structures. Kreimeyer (2009, pp.146ff.) provides a comprehensive set of metrics for the analysis of process structures. Freire and Alarcón (2002) and McManus (2005) adapt VSM metrics to design processes. Here, users often derive actions for improvement from generic principles, e.g., the 'flow'-principle in developing a future state VSM. The LPS analyzes process reliability by comparing planned task execution to actual task execution (Ballard 2000c, p. 1-6). Pall (2000, p.163) suggests a similar metric. Sosa et al. (2004, 2007) compare product architecture and organization architecture to find causes for misalignments. Cataldo et al. (2006) present a similar approach in the field of software development. Sosa (2008) presents a method for predicting communication when changing the product architecture. Kreimeyer et al. (2007) compare planned process structure to planned organization architecture. Chinowsky et al. (2011) compare actual and planned communication between people in the organization. Reichardt et al. (2012) present a method for

comparing actual and planned communication, partially based on Hickethier et al. (2011b, 2012).

(3) Feedback with Lessons Learned approaches, e.g., 'after action review' project post mortems, or post project reviews. These methods focus on learning through de-briefing, i.e., documenting experiences and lessons learned through workshops. Carrillo (2005) reviews and applies after action review to construction projects. Schindler and Eppler (2003) review several lessons learned approaches and present success factors for application on projects. Koners and Goffin (2007) analyze post-projects reviews for design projects. Lean construction practitioners often use the plus-delta-review for post project reviews (Howell and Macomber 2002).

3.2 Identification of Research Gap

3.2.1 Descriptive Study of Communication Structures in IPD Project Design Organizations

The research gap regarding communication structures of IPD projects focuses on the characteristics of actual communication in IPD project detailed design organizations. IPD proponents have argued that IPD projects act as a collective enterprise (Thomsen et al. 2010), implying that IPD-type contracts (1) establish organizational integration and (2) enable flexible organization (see subsection 2.3.2.2). Studies exist that prescribe how to achieve an integrated and flexible organization and present case studies in which flexibility fostered innovation (e.g., American Institute of Architects 2007; Matthews and Howell 2005; Thomsen et al. 2010). However, no descriptive study of IPD design organization exists that analyzes integration and flexibility from a communication structure perspective. Hence, the research gap regarding planning of communication structures in IPD projects pertains to whether integration and flexibility has been achieved. Part of the research gap is an analysis of the use of integrative mechanisms on IPD projects.

3.2.2 Planning of Communication Structures

The research gap regarding planning of communication structures focuses on learning about how to plan communication structures.

The presented methods for planning of communication structures focus on deducing structures from the product and process domains into the organization domain. Browning (2009) explains the importance of structural and non-structural mechanisms for integration of teams, and integration impacts communication between people. Design Process, organization architecture, and integrative mechanisms must consider uncertainty, because design is an inherently uncertain task.

The considerations regarding uncertainty demand iterative planning, checking of communication structures, and learning regarding lateral relations under dynamic uncertainty. Learning can be enhanced by improved transparency, regarding actual communication structures.

3.2.3 Improvement of Communication Structures

The research gap regarding improvement of communication structures focuses on self-reference of the AEC design system. The domains operating system and organization of the AEC design system are highly interdependent and a feedback loop shall connect these interdependent domains. The analysis of the point of departure concerning methods for improvement identified 3 types of methods for self-reference regarding actual structures:

- VSM-based methods,
- Last Planner System,
- Methods based on structural complexity.

The three following subsections will analyze these methods in detail and then outline the research gap.

3.2.3.1 VSM-based Methods

Freire and Alarcón (2002) apply VSM to design by capturing the current state of a design process. They identify waste and opportunities for improvement through a survey of employees working on the design process and they capture the design process on a low level of detail; activities are, e.g., "design", "review", and "release". Improvement focuses on reduction of inventories and increase of process flow. Freire and Alarcón (2002) do not capture design iterations.

McManus (2005) also captures the current state of design process with VSM. He includes iterations into the current state VSM modeling the structure of the design process with DSM. Improvement focuses on increased process flow through takt-time, line balancing, and streamlining of review processes.

Both methods capture actual information flow including processing times and waiting times. Both methods identify opportunities for improvement through the application of lean principles, e.g., flow and pull, and methods, e.g., takt-time and line balancing. Neither method compares actual information flow to planned information flow.

3.2.3.2 Last Planner System in Design

The LPS compares planned to actual task completion, based on the scientific experiment method. Implementation of LPS has led to gains in process reliability and productivity in design. One contributing factor is learning through analysis of work which was not completed as promised. Even though the LPS compares planned with actual task completion, the LPS can oversee opportunities for improvement due to the characteristics of information flow in design processes:

(1) Invisibility of information flow

The ability to learn is influenced by the visibility of the process at hand. In construction processes, it is easier to follow the flow of material along the process until the breakdown occurs than it is to follow information along the design process. Increased transparency of information flow in design can ease finding root-causes for deviation and thereby improve the ability to learn.

(2) Structure of information flow

The LPS achieves transparency on the structure of actual communication at discrete points in time, i.e., when a commitment is due. LPS can fail to find wasteful iteration between design tasks, if the task which was committed to has been completed in time but its execution triggered wasteful iteration. Also, the LPS can fail to standardize value-adding process structures, for example, if coordination between designers before completion of a specific task is not part of the process structure. In this case, a lack of standardization may lead to late rework of a task, when in some instances undocumented but necessary positive iteration had not occurred earlier in the process.

3.2.3.3 Structural Complexity-based Methods

In the field of DSM, Kreimeyer et al. (2007) compare the prescriptive design process model with the planned organization architecture to improve the fit between both. The focus on aligning planned communication structures neglects opportunities for improvement through analysis of actual communication structures.

In the field of SNA, Chinowsky et al. (2011) compare actual to planned communication between people in a design organization. They base the model of actual communication on a survey of people and they deduce planned communication from the network of planned tasks. They compare actual and planned communication between two people and find deviations between actual and planned communication. The comparison of communication between two people takes a transformation perspective, it does not include the analysis of integrative mechanisms, and it stops at finding deviation without researching related root-causes.

The comparison of actual and planned communication can be improved by (1) analyzing deviations from process flow perspective, and (2) by extending the analysis of planned communication with organization architecture and integrative mechanisms.

3.2.3.4 Summary of Research Gap

Current literature does not provide methods for comparison of patterns of actual and planned communication with the purpose of improvement based on the scientific method. To the knowledge of the author, DSM has not been applied to compare actual and planned communication. SNA applications have compared actual and planned communication, but they lack a focus on improvement. Several methods establish self-reference in the AEC design system, but no method exists that applies a comparison between patterns of actual and planned communication to trigger learning.

Clarkson and Eckert (2005, p.70) explain this research gap:

> "[...] hardly any company goes to the trouble of comparing the model with the process that actually exists. Process post mortems are rarely done, because everybody is busy moving onto the next project. While some main lesson might be learned, this is rarely about the process model itself."

Often, design processes are assumed to not be repeated due to the unique character of each design project. Therefore, the benefits of checking whether or not the process was conducted as planned may be considered minimal. However, opportunities to learn arise from comparing actual to planned information flow:

- learning about the process model at hand, i.e., improving the planned process, and learning about process modeling, i.e., improving modeling skills.
- learning about influences on actual information flow from inside the project system that steer actual information flow away from the planned structure, e.g., organization architecture and integrative mechanisms.
- learning about influences on actual information from the project environment that steer actual information flow away from the planned structure.

These opportunities to learn represent the motivation for filling the research gap. The research gap pertains to a method which compares actual and planned communication and can harvest these opportunities to learn. This method shall be rooted in (1) the flow perspective of the TFV-model, (2) the scientific experiment model, and (3) an open system perspective on the AEC design process.

Table 17 summarizes and structures the research gap into three parts and provides the related research questions, whose answers will be elaborated in this dissertation.

Table 17: Identified Research Gaps and related Research Questions

Research Gap	Research Questions
Lack of transparency regarding actual communication.	Q1. How can a design team efficiently achieve transparency of actual and planned communication in the detailed design phase of a construction project?
Improvement of self-reference of the project system.	Q2. How can the design team evaluate alignment of actual and planned communication? What are the metrics for evaluation?
Application of scientific experiment model in the analysis of differences between actual and planned communication.	Q3. How can the team use knowledge about misalignments between actual and planned communication to improve the design system structures continuously?

3.3 Requirements for Filling the Research Gap

3.3.1 Study of Communication Structures in IPD Project Design Organizations

Flexibility is a prerequisite for identified improvement to actually catch on. A rigid structure prohibits or impedes change, whereas a flexible structure fosters it. Hence, flexibility of communication structure is a prerequisite for successful implementation of goal-oriented change, i.e., change for the better.

The term IPD-projects describes a class of projects that employ IPD-type contracts. However, no two of these projects are the same. Hence, this research can only provide evidence that IPD-projects can have flexible and integrated communication structures. But this finding applies only to the project researched. Nevertheless, such a case study can help in drawing conclusions especially when it includes a description of applied integrative mechanisms.

3.3.2 Method for Improvement of Communication Structures Using Delta-Analysis

A comparison between actual and planned communication structures focuses on the identification of misalignments, or a delta between structures. Comparison cannot function without models of actual and planned communication structures. The purpose of comparison is to identify root-causes for misalignments. Comparison of structures is the starting point for root-cause analysis and learning with the method.

The method must consist of a (1) model which provides the theoretical underpinning for comparison of structures and (2) a procedure which considers the open-system perspective and the flow perspective during analysis.

Application of the method must be efficient. That is, the effect of learning must outweigh the effort for conducting the analysis. While it is almost impossible to monetize the identification of root-causes for deviations, it can be stated that resolution of the identified problems reduces waste. Hence, an evaluation of the method must consider its impact on the project as well as resources, and it must analyze the necessary skills of user of the method and possible barriers to implementation. The following questions regard impact and procedure (IP) of method:

- IP1. What are the qualitative impacts of application of the method on cost, quality and schedule?
- IP2. What resources are needed to implement the method?
- IP3. Who leads method implementation? What skills are necessary for implementation?
- IP4. What barriers to implementation of the method exist?

3.4 Summary

This chapter presented a framework for structuring the AEC design system. This chapter also analyzed existing methods for planning and improving the design system. Analysis of planning identified the research gap regarding a descriptive study of an IPD project design organization. The purpose of the study is to check whether the prescribed characteristics of integration and flexibility actually exist. Analysis of improvement highlighted the research gap regarding a method that compares actual to planned communication structures with the goal of identifying improvements based on the scientific method.

This dissertation focuses on two tasks to close the described research gaps:

(1) Exemplary proof of flexible and integrated communication structures in IPD projects. Chapter 4 presents case study (A), which analyzes integration and flexibility of an IPD-project based on the communication structure.

(2) Development and test of a method for improvement of communication structures using delta-analysis. Chapter 5 presents the method and chapter 6 presents two case studies (B1 and B2), which employ the method.

4 Case Study A – Social Network Analysis of Communication in an IPD Project Design Organization[25]

4.1 Introduction

A key principle in lean construction is to concurrently develop product and process during the design phase. This is enabled by bringing Last Planners from construction into the design phase while aiming to achieve a common understanding about the project early on between all involved parties. This approach increases the number of people involved during design, and thus increases the need for coordination. During the design phase, coordination means management of the information flow. To manage information flows, specifically on IPD projects, the team can apply specific mechanisms and roles, for example, cross-functional teams, cluster leaders, Chief Engineer position, collocation, Big Room, and Core Group.

Lean construction proponents claim that IPD projects achieve innovation and optimization across firm boundaries through integration and flexibility (see section 2.3.2.2; table 12). Relational contracts are the foundation of integration, which is amended with fitting organization and operating system (Howell et al. 2011; Thomsen et al. 2010).

The primary purpose of this chapter is (1) to document formal communication structures including organization architecture, integrative

[25] Parts of this section have been published in Hickethier et al. (2013).

mechanisms, and design process, and (2) to analyze organizational integration and flexibility with SNA by examining whether the mechanisms and roles prescribed in IPD-literature are actually in place. Analysis is followed by managerial recommendations and conclusions.

4.1.1 Characteristics of IPD Project Organizations

IPD-type projects differ in several regards from AEC projects that apply other kinds of commercial terms. This section describes the characteristics of organization, operating system, and commercial terms in IPD-type projects.

Projects are temporary social-technical systems, completed usually not by an individual, but by a group of people who must interact. This interaction is driven by the characteristics of the project delivery system, namely the 'project organization,' the 'operating system,' and the 'commercial terms' (Thomsen et al. 2010). Thus, project organizations cannot be analyzed independently from their context, namely operating system, and commercial terms (Howell et al. 2011).

Commercial Terms and specifically the relational contract terms used to define IPD projects promote collaboration between project members by including mechanisms such as pain-and-gain sharing, collective risk management, and contingency sharing. These mechanisms affect the relations between project members and promote strong collaboration. (Howell et al. 2011; Thomsen et al. 2010).

The operating system of IPD projects is based on the principle of reliable workflow (Howell et al. 2011). Key practices for increasing the reliability of information flow in design are, e.g., learning through PDCA thinking and root-cause analysis, look ahead planning with the LPS, VSM, and TVD. These practices build on small batches of information in design and a high frequency of information transfer.

Project organizations that follow an IPD agreement integrate owners, designers, and contractors. Contractors join the design team early and all partners work from a collocated office. Cross-functional teams consisting of individuals from the relevant companies find innovative and efficient solutions through their diverse set-up. An executive committee consisting of members from the involved companies manages the teams, makes decisions unanimously through consensus, and creates an open, collaborative culture. This model creates a "virtual company" (Thomsen et al. 2010) with members employed by their home companies but trusting each other strongly. The resulting collaboration fosters the behavior that the best qualified person does a job, regardless of their home company.

4.1.2 Case Study Description

Data was collected at the Van Ness and Geary Campus (VNCG) Hospital Project in San Francisco, California, USA, formerly known as Cathedral Hill Hospital (CHH) project. This project is well documented through prior research regarding:

- Operating System (Hamzeh et al. 2009; Lostuvali et al. 2012)
- Commercial Terms (Heidemann and Gehbauer 2010; Lichtig 2005)
- Project Organization (Hamzeh et al. 2009; Lostuvali et al. 2012)

The VNGC project applies a relational contract that falls into the category of IPD contracts, called the Integrated Form of Agreement (IFOA). The project members apply numerous lean principles and methodologies, among others: TVD, LPS, and A3 Reports. Project members are collocated in an office and operate in cross-functional teams, called 'Cluster Groups,' under the supervision of a Chief Engineer and an Executive Committee called 'Core Group.'

Data collection for case study A proceeded in two steps:

(1) Documentation of formal structures

The author joined the project team in the collocated office for 4 months to collect data regarding organization architecture, integrative mechanisms, and the design process.

(2) Analysis of informal structures

The author analyzed the informal organization structure based on a descriptive model of communication between people in the design organization. A survey provided the data for a model of the informal organization. SNA served to analyze the model, based on hypotheses derived from the related literature on lean design management.

4.2 Documentation of Formal Communication Structures at the VNGC Project

4.2.1 Scope of Project Phase

Documentation of formal and informal structures took place between July and October 2011. During that time, the design team worked on detailed design of the facility, specifically on design optimization and trade coordination. The project phase combines product design and production planning. It also details existing design with a constructability review that involves coordination between trades. Figure 34 shows the interconnectedness between design optimization and trade coordination.

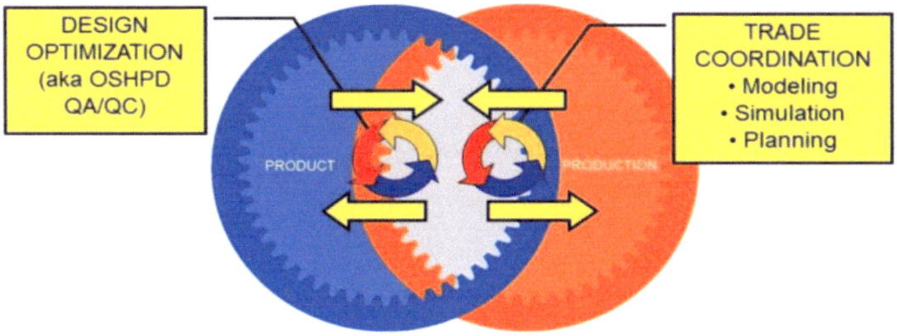

Figure 34: Scope of Project Phase (courtesy of Baris Lostuvali, VNGC)

4.2.2 Organization Architecture

The architecture of the design organization at the VNGC project follows the principles of the Toyota Product Development System as outlined in Morgan and Liker (2006). Important characteristics of the organization architecture at Toyota are 'Chief Engineer' and 'module development teams' (MDT). The MDTs develop subsystems of the product in-line with measurable goals that are agreed on with the Chief Engineer. The Chief Engineer and MDT leaders align all MDTs in their work (Morgan and Liker 2006, pp.131ff.). MDTs include experts, who come from different functional groups. Hence, the PD organization architecture at Toyota can be characterized as a matrix organization. The organization architecture follows the concept of 'concurrent engineering' (here called simultaneous engineering) by integrating production engineers during PD (Morgan and Liker 2006, pp.154f.).

A similar organization architecture was installed at the VNGC project. A Chief Engineer aligns cluster groups (comparable to MDTs). Each cluster group consists of people from the owner organization, designers, and contractors. Similarly to the organization architecture at Toyota a matrix organization unfolds, where home companies of the designers and builders substitute Toyota's functional departments. The structure of cluster groups centers around the project, and as such the organization can be characterized as a heavyweight project matrix organization (Ulrich and Eppinger 2004, pp.26f.).

However, in contrast to the heavyweight project matrix organization, cluster group leaders do not have formal authority nor are they involved in cluster members performance evaluations. This is in-line with Toyota's approach to the matrix organization (Morgan and Liker 2006, p.143) and it propels a leadership style in which people must be convinced with facts instead of authority. Leaders shall "lead as if they have no power"[26]. This mentality impacts project culture, where focus remains on the delivery of customer value. This focus is communicated and emphasized through the organization's vision, the five big ideas (Lichtig 2005), visualizations, project guides, and leadership, but also engrained in methods, e.g., TVD and the meetings and tools that establish TVD practice.

The organization architecture of the VNGC project changed as the project moved closer to construction. During the early stages of the detailed design phase (left hand side of figure 35) interdisciplinary cluster groups were responsible for different building systems, and these cluster groups coordinated through processes and meetings (Hamzeh et al. 2009). As the project moved along in detailed design to the subphase Design and Trade Integration Phase, the involvement of builders grew, communication focused more on the construction process, and the organization architecture changed. Groups were 're-chunked', by integrating designers more closely and defining a new group 'construction'. The three cluster groups – interior, structure, and technology – were temporarily stopped and met very infrequently. People were instead integrated with the MEP cluster group or the design cluster.

[26] Personal communication with David Thomack from VNGC.

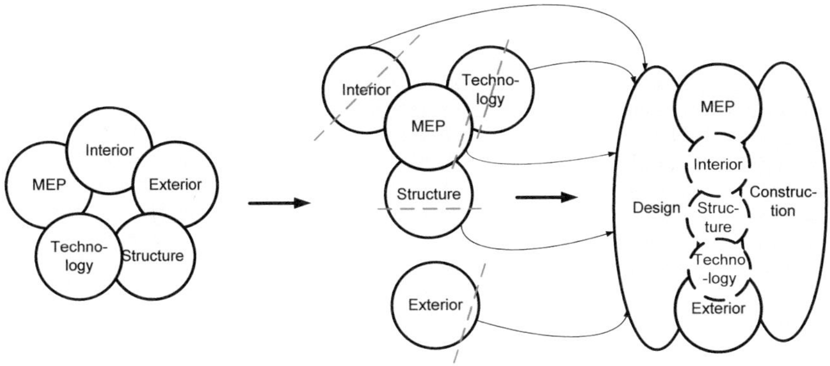

Figure 35: Change of formal Organization Architecture

4.2.3 Integrative Mechanisms

Integrative mechanisms bridge the gaps created by organization architecture. Each integrative mechanism has different characteristics that work toward the overall goal of integration. Table 18 provides a structured overview of the integrative mechanisms applied at VNGC. Table 8 explains each integrative mechanism (section 2.2.3.2).

Table 18: Integrative Mechanisms and their Application at VNGC

Integrative mechanisms (from (Browning 2009))	**Application at VNGC**
Improved information and communication technologies	BIM database, file server, email-lists
Training	Lean training, Study-action sessions, team-building events
Collocation	Collocated project office
Traditional meetings	Cluster Group meetings, Leadership meeting, TVD meeting
Town meetings	(n/a)
Manager mediation	Chief Engineer, Cluster Leaders
Participant mediation	Flexible organization enables people to act as coordinators when necessary
Interface management groups	Ad-hoc task forces for design issues that span cluster boundaries; Chief Engineer team
Standard processes	CBA, A3 Reports, Value Stream Mapping
Boundary objects	Share models of production and management through study action sessions.
Incentive systems	IFOA-contract
Shared interpretation of design problems	Set-based design
Shared knowledge	Reduced liability in IFOA incentivizes to share preliminary information.
Shared ontologies	Development of common language, e.g., CBA
Situation visibility	Visualization of budget, scope, current work, and improvement items (A3-reports)

Integrative mechanisms are also enablers for organizational flexibility. The concept of flexibility of organization architecture is based on individuals' behavior; flexibility means that people can change the project structure bottom-up. The project must provide the ability for people to change. Change consists of two steps:

(1) Awareness: realization that a different communication pattern might be better. For example, a team member realizes that he/she should switch from one group to another group or that he/she should attend a meeting of a different group to obtain needed information. Awareness necessitates the ability to obtain information quickly and to be able to draw the right conclusion based on personal knowledge.

(2) Action: the ability to quickly integrate into a different team, group, or process in order to make the change of behavior successful.

The integrative mechanisms presented in table 20 serve both steps. Some focus on the distribution of information and others on building a common ground through which people interact, e.g., common language and vocabulary and a shared understanding of the project.

4.2.3.1 Integration by Achieving Awareness through Communication

All of these integrative mechanisms impact not only the pattern of communication but also the behavior of every individual person. Several of these integrative mechanisms correspond to communication channels. Communication channels have several characteristics, two of which are reach and mode of information transfer:

- Reach of communication channels

Some communication channels, e.g., IT-servers, are equally accessible to everyone who has access, i.e., the reach of these communication channels is equal for all people. Other communication channels impose a structure of information flow in which different people have different reach, e.g., emails send through distribution lists only reach members of that list.

- Mode of transfer of communication channels

Distribution of information can follow two basic approaches: push and pull. Push broadcasting is supplier driven; he/she broadcasts information to people he/she deems suitable. Information pull is customer driven; he/she collects necessary information as needed. Morgan and Liker (2006, p.95f.) highlight the importance of information pull at Toyota. Application of information pull necessitates that people have access to relevant design data and to people carrying information, including leadership positions such as the chief engineer. At Toyota it is the job of the Chief Engineer and of MDT leaders to achieve coordination and alignment with other MDTs.

VNGC implements these prerequisites for successful information pull through collocation, open servers and 3D-models, and a project culture that makes leadership approachable. Also, cluster leaders are responsible for coordination with other clusters.

Morgan and Liker (2006, p.97) describe the limitations of information pull, noting that "the level of the routine processes in manufacturing is not possible within product development". Uncertainty impacts design processes and makes information distribution more probabilistic than material flow in production. Hence, some information may be misrouted which can have two effects:

- a person, who actually needs a piece of information, does not receive it. The unawareness about a necessary input for a task can lead to poor quality of results and cause wasteful rework;
- a person, who does not actually need a piece of information, does receive it. Unnecessary information can spur creativity or cause information overload.

Here, conflict surfaces between effectiveness and efficiency of information flow (see section 2.2.3.3).

Table 19 presents communication channels at VNGC with their mode of information transfer, reach, and management approach.

Table 19: Communication Channels at VNGC

Integrative Mechanism	**Communication Channel**	**Mode of information transfer**	**Reach**	**Management approach**
Collocation	Face-to-Face	Pull	Related to communication – physical distance curve (Allen 1977, p.241)	Collocation – seating Chart managed by Director of Continuous Improvement
Improved information and communication technologies	Direct Email / Telephone	Push /Pull	All people (with access rights)	Address directory accessible for all people at the project
Improved information and communication technologies	Email Lists	Push	People who are on each list	Emailing Lists managed by IT Support. People can request to be added and taken off
Traditional meetings	Meetings	Push (presentation) / Pull (questions)	Attendees of meeting, often related to cluster group structure, often managers of other cluster groups and chief engineer attend (manager	Standardized weekly meeting schedule
Improved information and communication technologies	3D Model	Pull	All people (with access rights)	3D model accessible on server; access rights set by BIM administrator
Improved information and communication technologies	Document Server	Pull	All people (with access rights)	Documents accessible on server, access rights set by IT support
Situation visibility	Visual Management including transparency of process model	Pull	Often related to communication – physical distance curve, because visualization are located close to owner of visualization (Allen 1977, p.241)	Process model visualized in Big Room and accessible on server. Visualization of, e.g., processes and building systems, located close to owning cluster groups

Weak ties (see section 2.2.3.2) give access to new information, which is more probable to spur innovation. Figure 36 shows the distribution of different tie strengths of collocated and non-collocated people at the VNGC project. This analysis is based on the survey presented in section 4.3.3: 54 survey participants worked from the collocated office and 14 worked from other office locations. Figure 36 shows that people working from the collocated office have almost double as many monthly and weekly information exchanges with other people on the project than people who work remotely.

Quick access to people in the collocated office and aligned interests through the IPD contract support the development of weak ties. Also, the collocated office can be seen as mechanism for building trust: collocation reduces the power of information brokers, because other parts of the project organization are within walking distance. Transparency about the status of work throughout the project organization reduces opportunities for information brokers to abuse their position for their own benefit. Consequently, it can be assumed that people on the project are willing to trust each other faster than on people on projects without an IPD-type contract.

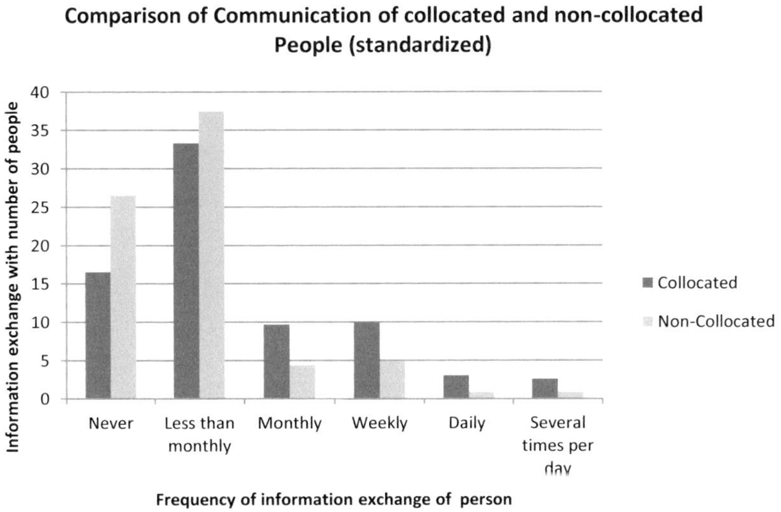

Figure 36: Communication of collocated vs. non-collocated People

4.2.3.2 Integration by Enabling Quick Action

At least five integrative mechanisms contribute to the ability to integrate quickly and enable people to take action effectively:

- standardization of processes, such as CBA, A3-Reports, and Value Stream Mapping, and the related training for people in the organization,
- boundary objects,
- incentive systems,
- shared interpretation of design problems,
- development of shared ontologies through a common language among project participants.

4.2.4 Design Process

Figure 37 shows the structure of the task network at the VNGC project. The task network can also be characterized as the network of commitments between people working on the project. Hamzeh et al. (2009) describe the planning process that is based on the Last-Planner-System: phase planning and look-ahead planning take place inside Cluster Groups. Tasks which span across Cluster Groups are discussed during special meetings.

The structure in figure 37 is based on the task network from SPS-Software and the process maps used in the MEP Cluster Group on October 18, 2011. The network consists of several disconnected bodies of tasks and these bodies align to some extent with the cluster group structure of the project. Further research is necessary to understand the reasons that lead to these disconnected groups of tasks.

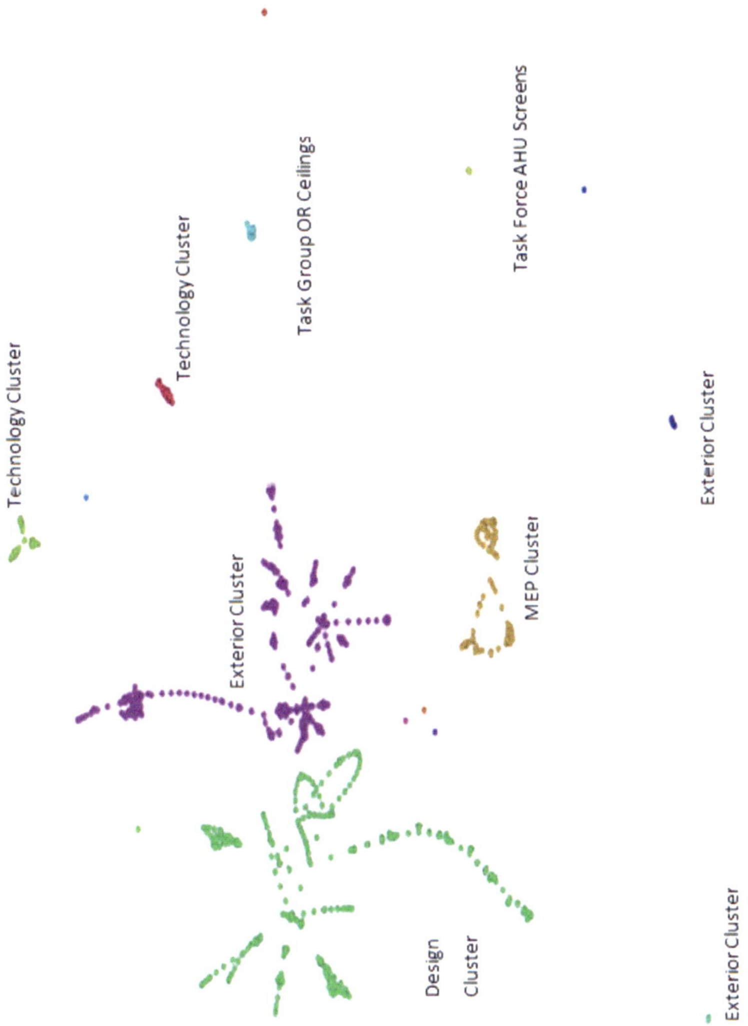

Figure 37: Structure of Tasks at VNGC project on October 18, 2011

4.3 Social Network Analysis of Communication in the Informal Design Organization

This section is structured as follows: first the author reviews the literature regarding SNA, characteristics of communication in design organizations, and specifics of IPD projects. Second, the author analyzes three IPD-specific coordination mechanisms and roles and presents SNA indices for their assessment with hypotheses. Third, the author presents the case study and the research methodology. Fourth, the author presents the findings based on the data gained in this case study A. Fifth, the author presents managerial recommendations for coordination of IPD-projects. Sixth and last, the author closes the section with conclusions and recommendations for future work.

4.3.1 Social Network Analysis

Moreno (1934) introduced Social Network Analysis (SNA) by using sociograms, which are formal representations of social relationships between people visualized through graphs. The sum of relationships between two actors constitutes the connection, or tie, between them, and the sum of ties between all actors constitutes the social network (Wasserman and Faust 1994). The goal of SNA is to build the social network empirically based on observed interaction. Based on these interactions the informal structure of the network unfolds. This approach differs from the defining the formal network structure prior to interactions, for example by creating the organizational structure of a company or project.

Braha and Bar-Yam (2004) show that the connectedness of tasks in product development projects follows a power law distribution, i.e., few tasks are highly connected with other tasks, while many tasks are sparsely connected. This network characteristic implies that connectedness between people in the network is not evenly distributed. Instead, few very well-connected people control the information flow within the organization. These people are critical for the success of the project, because their position within the network gives power and influence. Social Network Analysis (SNA) has been successfully applied to identify these critical people based on indices, such as centrality, betweenness, and clustering.

Using SNA in a case study, this research applies these indices to analyze an IPD-project's design organization. The goal of this research is to evaluate the use of aforementioned IPD-specific mechanisms and roles. Specifically, the author tests hypotheses regarding cross-functional teams, and the roles of cluster leaders and the chief engineer.

4.3.2 Network Properties and Hypotheses

Ties between actors can be defined as existing vs. non-existing, or each tie can receive a value to reflect a weight. SNA devotes special attention to the role of weak ties. Granovetter (1973) sees infrequent and distant relationships as sources for diverse information through remote people, who are more probable to have new knowledge.

Wasserman and Faust (1994) list a number of network properties with corresponding indices to assess a social network. This research focuses on centrality and component aspects of the network. The following paragraphs explain how these aspects relate to coordination mechanisms and roles in design organizations.

4.3.2.1 Centrality Aspects of a Network

An individual is called 'central' when they are connected to a large number of other people in the network, either directly or indirectly. Wasserman and Faust (1994, p.178) describe centrality using three different indices: (1) degree centrality, (2) closeness centrality, and (3) betweenness centrality. In this section the author applies indices

(1) and (3). Figure 13 (section 2.1.4.4) illustrates individuals with respective centralities.

An individual with a high degree centrality is very communicative and directly relates to a large number of other people in the network. Their centrality presumably corresponds to the power and influence they have in the network. Leaders of cross-functional teams are highly connected to the members of their team, but also coordinate with leaders of other teams. Thus, the author proposes hypothesis 1: leaders of cross-functional teams have a high degree centrality.

A person with high betweenness centrality is in a brokerage position and can exercise strong power and influence in the organization. In design organizations he/she is a broker for information and acts as a gatekeeper or mediator between otherwise disconnected parts of the network. Burt (2004) claims that a person in this position on average has more creative ideas than other people have, and their ideas are more likely to be accepted by others in the network.

The Chief Engineer coordinates work between cluster groups and, while not having formal authority, he/she is highly respected by all members of the project team, i.e., he/she has a very powerful position within the organization (Morgan and Liker 2006, p.132). Thus, the author proposes hypothesis 2: the Chief Engineer has high betweenness centrality.

4.3.2.2 Component Aspects of a Network - Clustering

Networks can be segmented into clusters. People inside the cluster are highly connected to each other but sparsely connected to people outside the cluster. In a design organization, such clusters represent teams, in which people frequently exchange information with each other while they do less with people outside their team. Thus, clustering of design organization reveals the structure of collaboration, i.e., how the people structure themselves within the informal organization.

IPD projects apply the coordination mechanism of cross-functional teams. This structure breaks the traditional three-silo-structure between owner, designer, and contractor, thus enabling global optimization of the design through integration of requirements from all three perspectives (Thomsen et al. 2010). Thus, the author proposes hypothesis 3: Clusters of the informal IPD organization consist of owners, designers, and contractors.

4.3.3 Research Methodology

The author conducted a survey on communication between people on the project team. Through the survey, each person could indicate the level of information received from and sent to others in the office. The survey focused on a three-month period and people were instructed, through prior team presentation and in the survey, to only register technical communication in the survey. Technical communication was explained as 'giving you the information you need to complete the work at hand.' People were instructed to consider all available channels of communication, e.g., face-to-face, email, telephone. Possible levels for information flow were 'never,' 'less than once per month,' 'monthly,' 'weekly,' 'daily,' and 'several times per day.' The author collected data regarding the information flow between 99 people in the design organization. 68 people successfully completed the survey on the website www.surveymonkey.com. Survey participants indicated the information exchange between themselves and 75 people on the project. To increase the utility of the data gathered, the author combined the *receive* and *send* perspectives into a combined map of communication. For this transformation, the author only included people that either (1) completed the survey, or (2) were listed in the survey itself. This resulted in a total of 99 people for the model.

Based on the information gathered through the survey, the author built a Social Network Model. People are represented as nodes. Communication between them is shown through weighted edges between the nodes.

Table 20 shows the translation from levels of information exchange into weighting of edges. The Social Network Model combines the *send* and the *receive* perspectives of information exchange, and these two perspectives do not completely align due to mismatched interactions. The *receive* perspective denotes 5% more interactions than the *send* perspective. The author conducted a sensitivity analysis based on three social network models:

- If in conflict, the higher of Send and Receive value is correct.
- If in conflict, the lower of Send and Receive value is correct.
- If in conflict, the average of Send and Receive value is correct.

Sensitivity analysis yielded no significant differences in centrality and component aspects of the three models (see appendix B). In-line with other researchers (e.g., Eppinger and Browning 2012, p.87) the author assumed that the higher of the two levels of information flow is correct.

Table 20: Weighting of Information Exchange for SNA

Level of Information Exchange	**Weighting of Edge**	**Rationale**
Never	0	-
Less than once per month	1	Max. once every 2 months
Monthly	2	[scale factor]
Weekly	9	4,5 weeks / month
Daily	45	5 days / week
Several times per day	90	At least twice per day

The author analyzed the resulting weighted social network model with the software Gephi (Bastian et al. 2009). The model used for cluster analysis with Gephi only considers weekly, daily, and several times per day levels of information flow. This filtering is based on the assumption that members of the same cluster communicate at least weekly. The model of weighted information flows represents integration of people into the design organization and it enables analysis of peoples' informal role within the organization.

4.3.4 Results and Findings

Distribution of connectedness between people in the design organization shows a pattern similar to the findings of Braha and Bar-Yam (2004): a large number of people exchange relatively little information with others in the organization (left side of figure 38), whereas a small number of people act as information hubs transferring large amounts of information (right side of figure 38). One may assume that information transfers between people on an IPD project are evenly distributed for two reasons: (1) the IPD contract fosters trust between all members of the organization, and (2) the workplace enables easy access to all people on the project. However, the analysis shows the existence of information leaders, who are highly influential in the project organization.

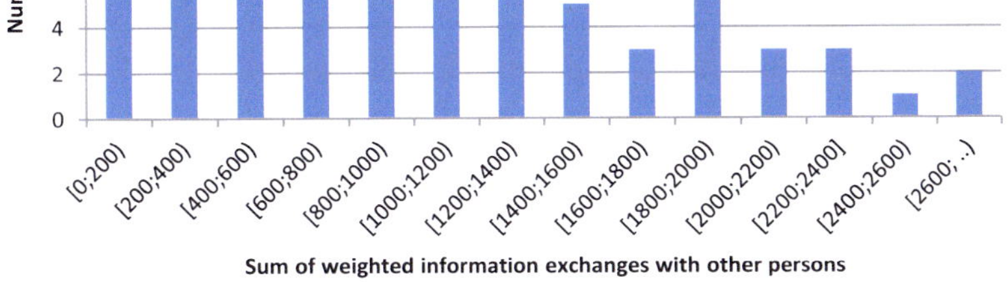

Figure 38: Weighted Degree Distribution of Information Exchanges

The cumulative weighted degree distribution (figure 39) shows similarities to a power-law distribution. This characteristic indicates organization robustness. Few people are highly active information hubs, so organization breakdowns due to random effects (e.g., illness of a person) are not probable. However, the organization is at risk for targeted attacks at highly active information hubs (see section 2.1.4.4).

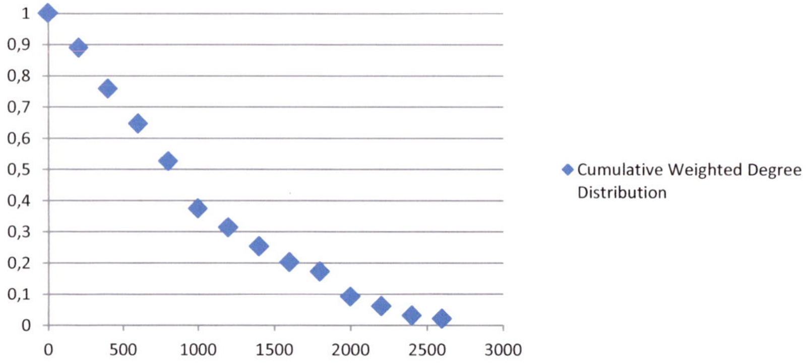

Figure 39: Cumulative Weighted Degree Distribution of Information Exchanges

Next, the author analyzed the mechanisms of information exchange within the design organization using degree centrality and betweenness centrality. Table 21 shows the highest ranking people, for weighted degree centrality and betweenness centrality. Numbers following roles, e.g., 'Electr. Designer', stem from anonymizing the data-set und represent a sequential numbering of people with the same role.

Table 21: People with 10 highest respective Centrality Indices in descending Order

Weighted Degree Centrality	**Betweenness Centrality**
PM Mech. Plum. Contractor	GC - Chief Engineer / Cluster Leader 4
Mech. Plum. Contractor 4	Owners Rep Core Group
Electr. Designer 7	GC - Chief Engineer Staff
GC - Chief Engineer / Cluster Leader 4	GC - Cluster Leader 1
GC BIM Expert - Cluster 3	GC - Cluster Leader 3
GC - Cluster Leader 3	GC - BIM Expert Cluster 4
GC - Cluster Leader 2	PM Mech. Plum. Contractor
Arch 10	GC 7
Electr. Contractor 2	GC - Cluster Leader 2
Arch 2	GC 5

Data supports hypothesis 1, 'leaders of cross-functional teams have a high degree centrality.' Three of the four leaders of the cluster groups (at the VNGC project called cluster leaders) lie within the 10 people with the highest weighted degree centrality. In this case study, the Chief Engineer has a double role, since he acts also as Cluster Leader four.

Data also supports hypothesis 2, 'the Chief Engineer has high betweenness centrality.' The Chief Engineer lies within the 10 people with the highest betweenness centrality.

Table 21 also shows that information leaders outside the assigned coordination staff exist, for example 'PM Mech. Plum. Contractor,' 'Mech. Plum. Contractor 4,' and 'Electr. Designer 7'. This finding highlights that IPD projects encourage people to do what is necessary to make the project successful, regardless of their formal role.

Flexibility is a structural characteristic of IPD projects (see section 2.3.2.2, table 12). Analysis of degree centralities shows that people become information hubs, even when this kind of coordination is not part of their formal job description. This characteristic is an indicator for the existence of structural flexibility in the design organization, because people can reach influential position (according to their position in the social network) even though this influence is not part of their formal role. Incentives and bottom-up management achieve a flexible organization architecture, where every person can influence communication patterns and become an information hub.

Figure 40 shows a force-directed graph of the design organization (labels represent people, arrows represent communication between them). In a force-directed graph, connections between a pair of nodes can be seen as springs that try to pull the pair closer together. The algorithms used to lay out this graph (namely Gephi's 'Force-Atlas 2' and 'Label Adjust') minimized the sum of spreads of all springs in the graph. These algorithms considered only information exchange levels 'weekly,' 'daily,' and 'several times per day,' and accordingly figure 40 shows only these levels.

Data partially supports hypothesis 3 'Clusters of the informal IPD organization consist of owners, designers, and contractors.' Figure 40 shows the four distinct clusters in different colors as found by Gephi's clustering algorithm. Designers and contractors highly interact inside these four clusters; however three of the four clusters do not include owner representatives.

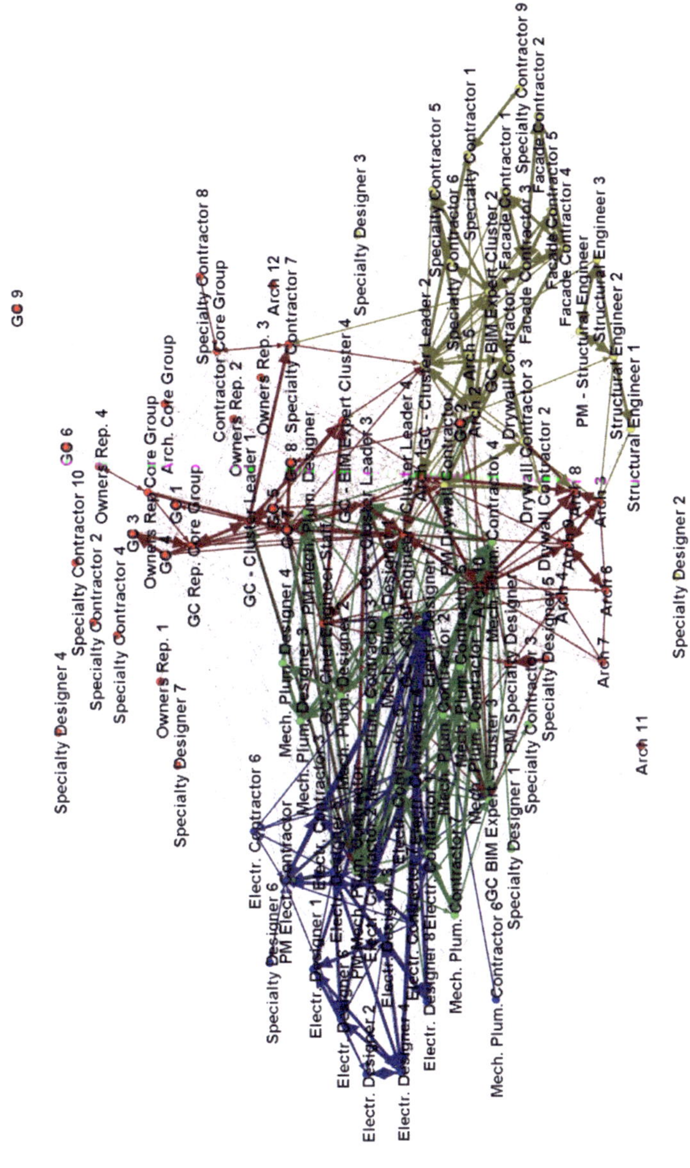

Figure 40: Force-directed Graph of Project Team - Colors indicate Clusters as found through Clustering Algorithm

Figure 41 shows the organization architecture in two organization DSMs. The left hand DSM is sorted by company, so highly connected groups of people from the same company become visible along the diagonal. The right hand DSM is sorted by cluster groups, so that highly connected groups of people from the same cluster group become visible along the diagonal. The right hand DSM also marks cluster leaders (in green) and the Chief Engineer (in yellow). Larger depictions of both DSMs can be found in appendix A. Both perspectives on the project organization show strong interaction (1) between people from the same company, and (2) between people from the same cluster group. The cluster group perspective also shows the interface function of the design cluster, mainly through people from the GC and the architecture firm: the design cluster acts as an interface between the other clusters.

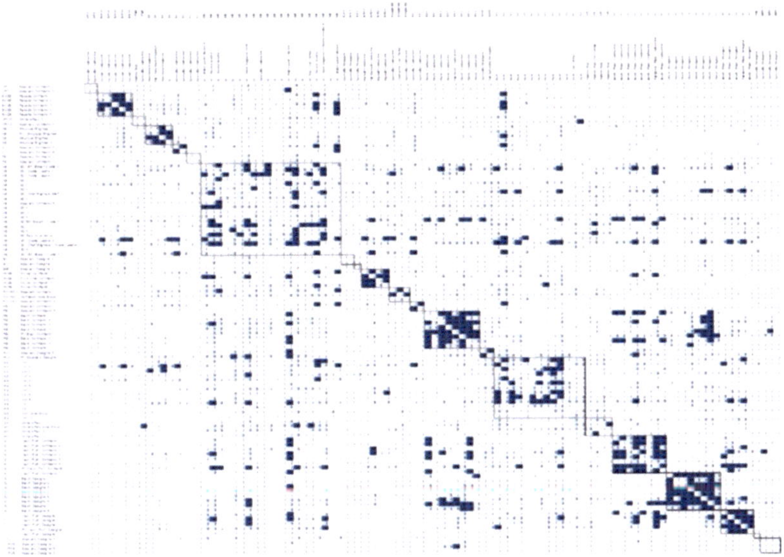

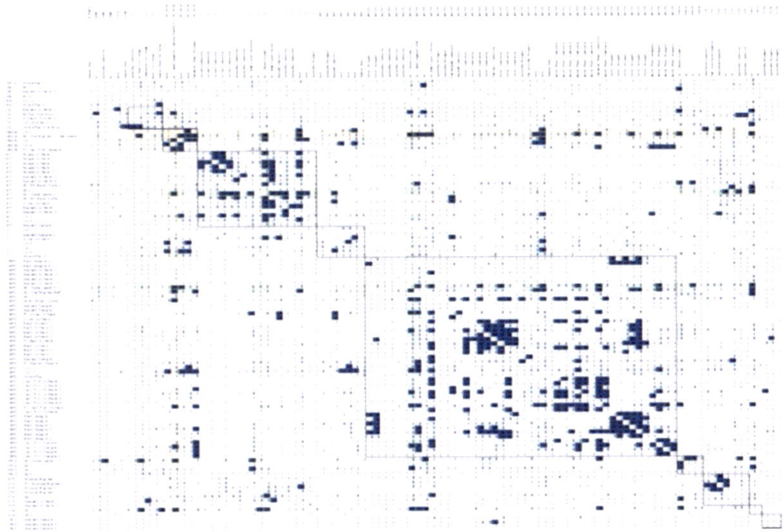

Figure 41: Information Exchange between People sorted by Company (top) and sorted by Cluster Group (bottom); see appendix A for larger figures

4.4 Managerial Recommendations

As shown, this IPD project encourages people to get involved for the benefit of the project. People, who see themselves as capable, coordinate work between others, regardless of their formal role. This bottom-up approach to managing communication is beneficial for coordination efficiency. However, if people who coordinate communication are not qualified for the job, coordination may not be effective.

For example, BIM Experts have an important role during detailed design. The high centrality of BIM Experts in Table 21 shows their importance in the design organization. Not only do they coordinate between people within their own clusters, they also coordinate between clusters. This job increases the requirements on the role: in order to recognize potential for innovation and savings, BIM experts need expertise in building systems and technology on top of their expertise in BIM.

The author recommends that such information leaders in the informal organization be identified through SNA, and that those people be trained to qualify for the job of coordinating teams. Information leaders will probably change during the different phases of a project, so the author recommends that the search for information leaders be repeated.

Modeling and analysis of the social network can serve as a method for checking whether formal roles align with the informal organization, as demonstrated in this research.

4.5 Critical Review

Survey data served to model the informal organization. The data-set is incomplete, because not all project members completed the survey. That is, survey participants were asked to denote information exchange to and from 75 people. 68 people completed the survey, but the group of people noted in the survey and the group of people who completed the survey overlap only partially. The partial overlap between both groups results in an overall number of 99 people, for whom data regarding information exchange exists.

The survey collected two perspectives of information exchange per person: (1) what information is being received and (2) what data is being sent? These two perspectives generate a redundant model, because the 'send' perspective of one person is the 'receive' perspective of another person. This redundancy made possible a sensitivity analysis of the model and three different scenarios were compared (see appendix B). Sensitivity analysis yielded similar cluster structures and degree centralities in all three scenarios for 'key people' of this case study. 'Key people' refers to people who are included in a hypothesis of the case study, i.e., Chief Engineer and Cluster Leaders. Thus, even though the data-set of information exchange is incomplete, it is applicable towards the purpose of this research.

It should be stressed that significance of the analysis as conducted has limitations. These limitations are (1) that the case study analyzed only one project during a three-month period, and (2) that the case study analysis included all technical communication without further reviewing quality of information being communicated. For example, helping a less experienced designer includes per definition technical communication, assuming that the designer needs this technical information exchange to do his/her job properly. More frequent information exchange with a less experienced designer may lead to similar results as less frequent information exchange with a more experienced designer. Such differences in quality of communication were not included in the survey.

In the context of this case study, results regarding integration and flexibility of the formal organization were particularly interesting. Cluster analysis revealed that designers and builders interact within clusters. This finding is limited to the three-month period of the survey.

The research finding of flexibility is limited for two reasons: (1) currently no frame of reference exists regarding degree distributions in IPD-type projects and what roles have high degree centralities. Also, in non-IPD projects people could have high degree centralities without having a role that includes coordination work. (2) Flexibility is a dynamic characteristic of an organization and therefore it is better observed through changes of the organization architecture over time. This case study examined only one data-point in time which represents a three-month period, and therefore significance of the case study in terms of assessment of flexibility is limited. Longitudinal studies of communication would have higher significance in terms of the assessment of flexibility.

4.6 Summary

At the risk of over-generalizing from the set of data collected on VNGC, the author draws the following conclusions:

IPD practices promote an increase in the number of people involved in design, as compared to traditional projects (Thomsen et al. 2010, p.11). Thus, IPD increases the need for coordination of the larger design team. Collected data shows that the distribution of information exchange between people involved during design is uneven: many people exchange information sparsely, while a few individuals act as information hubs between separate parts of the network.

IPD proposes that owners, designer, and builders interact during the design phase of a project (section 2.3.2.2, table 12). Cluster analysis shows that designers and builders have strong interactions, because they mix in identified clusters of the social network. But all owner representatives are grouped in cluster of the social network, which suggests that interaction between (1) owners and (2) designers and builders is weak (as compared to interaction between designers and builders). This case study took part during the detailed design phase of the project. Future research is necessary regarding degrees of owner integration during different design phases. Nevertheless, the traditional silo-structure does not exist in this project, and instead the organization can be considered as integrated.

Degree distributions show that few people communicate often, and many people communicate seldom. People with a high degree centrality are associated with coordinating roles in the organization. Analysis of people with high degree centralities shows, that some of these people undertake this coordinating role, even though it is not part of their formal job description. This finding serves as evidence for flexibility of the organization architecture.

IPD projects run the risks of missing opportunities for innovation and cost savings, when people gain influential positions in the informal organization, without having the appropriate skills for coordinating others. SNA is a tool to identify influential people based on their communication patterns, so that they may gain skills to better fill this informal role.

Further research is necessary regarding the need for owner involvement, specifically regarding the frequency of interaction with designers and builders but also regarding the frequency of coordination between owner representatives. The questions "why is information exchange unevenly distributed on this IPD-project?" and "are other distributions of information exchange beneficial?" remain for future research.

5 Method for Improvement of Communication Structures Using Delta-Analysis[27]

5.1 Placement of Method in the Context of Organization Design

This chapter focuses on the development and description of a method. The method belongs to the field of organization design in AEC design projects. The objective of organization design is to divide a task into subtasks, assign these subtasks to people or teams, and then to integrate people or teams into an organization (e.g., Lawrence and Lorsch 1967). Integration refers to bridging the artificial gaps in the organization, which stem from partitioning. Integration differs from coordination. The need for integration depends on the dependencies between tasks and on uncertainty (Worren 2012, pp.168ff.).

Organization design establishes the organization architecture which includes at least two types of relations (Eppinger and Browning 2012, pp.80f.; Ulrich and Eppinger 2004, p.23): (1) reporting relations, which are mostly vertically arranged, and (2) lateral relations, which are mostly horizontally arranged. The method presented in this chapter focuses on improvement of lateral relationships.

This dissertation defines improvement based on the scientific method, and system self-reference through feedback is a critical component of the scientific method. Section 3.1.4.2 identified three classes of methods for feedback: methods providing feedback regarding performance,

[27] Parts of this section have been published in Hickethier et al. (2011, 2012b).

methods providing feedback regarding actual structures, and lessons learned approaches. The method presented in this chapter is part of the class of methods which provide feedback regarding actual structures.

Figure 42 builds upon the model for planning and improvement of communication structures (figure 32, section 3.1.3). The method described in this chapter relates to step 7 "evaluate status".

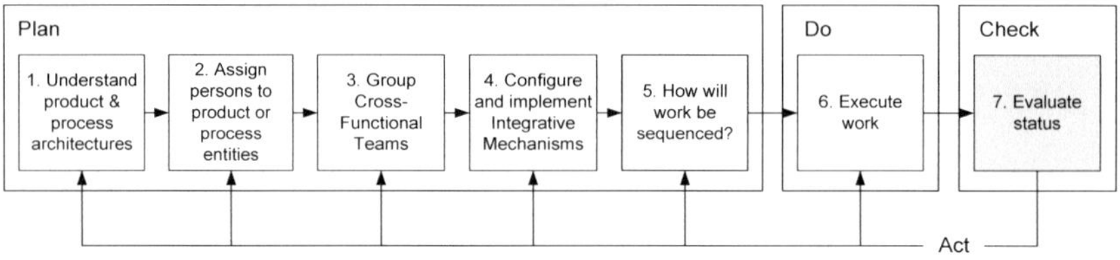

Figure 42: Placement of Method in the Context of the Model for Planning and Improvement of Communication Structures

5.2 Goal and Requirements of Method

The research gap comprises a method that compares actual to planned communication. The method is rooted in an open-systems view of the design process and the goal of the method is learning about communication structures.

The method elaborated in this dissertation checks misalignments between actual and planned communication. It is important to mention that the goal of the method is not to determine, which of the two perspectives on communication is correct. Also, it shall be noted that misalignments are not necessarily bad or should be avoided at all cost.

Misalignments can stem from several sources, for example, from falsely defined communication structures. In such a case, learning from the deviation between actual and planned communication structure can be beneficial for improving the project organization. Therefore, the method elaborated in this dissertation, applies the principle of investigation (section 2.3.2.1). The purpose of investigation is to find the root-cause for a deviation. Based on the identified root-cause actions for improvement can be taken. Thereby the method supports the development of a learning organization.

5.3 Modeling of the Method

5.3.1 Models of Communication and Information Flow

Communication and information are both part of this method. However, models of communication and information flow have different structural characteristics. Communication takes place between people and this dissertation follows Flores's (1981) LAP for modeling communication. Figure 43 shows communication between people in the organization domain based on LAP; it depicts the successful fulfillment of request at first try. In this example, communication consists of four basic parts which point both ways between person A and B. People communicate bi-directional, thus a model of communication can assume communication as undirected between people.

Information flow takes place between tasks of a process. Figure 43 shows an example in which information flow is the output of task A and the input of task B. The example shows the directness of information flow.

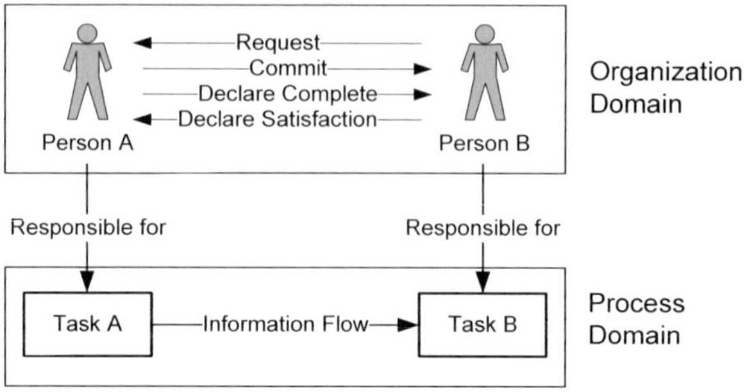

Figure 43: Directedness of Communication and Information Flow

In this dissertation the goal of modeling is to reveal communication between designers and then find differences between actual and planned communication. Comparison of structures necessitates models of structures. Usually, models of communication between people are not directly available and capturing communication takes considerable effort. Thus, the method uses indicators for modeling communication. Indicators are comparable to proxies. They may not capture all communication, but establish a meaningful model for the purpose of improvement of communication structures.

All indicators for communication are deduced to the organization domain: the author works with the assumption that designers can relate better to their communication with other designers, than they can relate to the abstract exchange of information between tasks of a process map. Root-cause analysis becomes more tangible for designers, when analyzing communication between their peers. Thus, the author defines the organization domain as the base for comparison of communication structures. Responsibilities of people for entities of the indicator domains are used to deduce relations between people in the organization domain; this mapping can be interpreted as an affiliation matrix. Figure 44 exemplarily shows the logic for deducing relations between people from indicators for actual and planned communication.

- Model of actual communication called the 'as-is' perspective.

Descriptive models of information flow, e.g., event logs in IT systems, serve as indicators for modeling of the 'as-is' perspective of information flow. Event logs in IT Systems can connect two people without denoting a direction of the relation, e.g., the event involves both people without documenting further details. Hence, figure 44 shows a bi-directional arrow between conducted task 1 and conducted task 2 (represented as beams 1 and 2 on the lower left hand side).

- Model of planned communication called the 'should' perspective

Prescriptive models of information flow, e.g., process maps, serve as indicators for modeling the 'should' perspective of communication. Process maps often document directed input-output relations between tasks, hence figure 44 shows a directed arrow between planned task 1 and planned task 2. The adjective 'planned' does not imply that planned communication is always the right way to operate. The 'should' perspective can be wrongly defined, e.g., incomplete, and therefore in need of improvement.

Both, 'should' and 'as-is' perspective can consist of one or more datasets. Figure 44 shows an example of the deduction of 'should' and 'as-is' communication from a prescriptive model and a descriptive model of information flow.

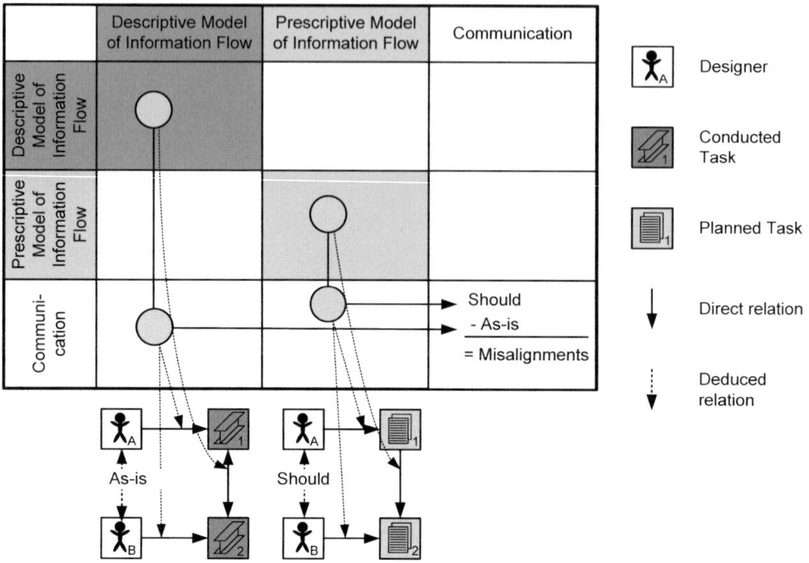

Figure 44: MDM Model of Combination of Descriptive and Prescriptive Process Models

5.3.2 Set Theoretical Model of Communication

The set theoretical model serves as the foundation for computing the delta between 'should' and 'as-is' perspectives on communication. Documentation of actual communication is time-consuming and sometimes even infeasible, because information flow in design is often invisible. Indicators can be used to approximate communication.

The model consists of three sets: indicator set 'should', indicator set 'as-is', and organization set. In the first step these sets are mapped onto themselves. The mapping establishes three square matrices which enable modeling of relations between the entities of each set.

$\text{Indicator Set}_{\text{Should}}$ $\quad I \in \{1, \dots, n\}, n \in N$

$\text{Indicator Set}_{\text{As-Is}}$ $\quad P \in \{1, \dots, m\}, m \in N$

Organisation Set $\quad O \in \{1, \dots, o\}, o \in N$

i – row, j-column

$\text{Indicator}_{\text{Communication Should}}$

$$DSM_{\text{Indicator,should}}: \{1, \dots, n\} \times \{1, \dots, n\}, \quad (i,j) \to i_{i,j}, \quad i_{i,j} \in \{0,1\}$$

$\text{Indicator}_{\text{Communication As-Is}}$

$$DSM_{\text{Indicator,as-Is}}: \{1, \dots, m\} \times \{1, \dots, m\}, \quad (i,j) \to p_{i,j}, \quad p_{i,j} \in \{0,1\}$$

Organisation

$$DSM_{\text{Communication}}: \{1, \dots, o\} \times \{1, \dots, o\}, \quad (i,j) \to o_{i,j}, \quad o_{i,j} \in \{0,1\}$$

The next step maps indicator set 'should' and indicator set 'as-is' onto the organization set, which includes people or teams, i.e., entities or a subset of entities from the organization domain. This mapping establishes a 'responsible for' relationship between people or teams and tasks.

$\text{Responsibilites}_{\text{Indicator Should}}$

$$DMM_{\text{Indicator,should}}: \{1, \dots, o\} \times \{1, \dots, n\}, \quad (i,j) \to s_{i,j}, \quad s_{i,j} \in \{0,1\}$$

$\text{Responsibilites}_{\text{Indicator As-Is}}$

$$DMM_{\text{Indicator,as-Is}}: \{1, \dots, o\} \times \{1, \dots, m\}, \quad (i,j) \to t_{i,j}, \quad t_{i,j} \in \{0,1\}$$

The next step calculates two matrices: Communication 'should' by multiplying the indicator matrix 'should' with the responsibilities matrix and communication 'as-is' by multiplying the indicator matrix 'as-is' with the responsibilities matrix.

$$f: DSM_{Should} \rightarrow DSM_{Communication} = DMM_{Should} \times DSM_{Should} \times DMM^{T}_{Should} = DSM_{Communication,Should}$$

$$f: DSM_{As\text{-}Is} \rightarrow DSM_{Communication} = DMM_{As\text{-}Is} \times DSM_{As\text{-}Is} \times DMM^{T}_{As\text{-}Is} = DSM_{Communication,As\text{-}Is}$$

Multiplying the matrices aggregates information flows between tasks into communication between people. Several information flows between people can be merged into communication, which is bidirectional (figure 45).

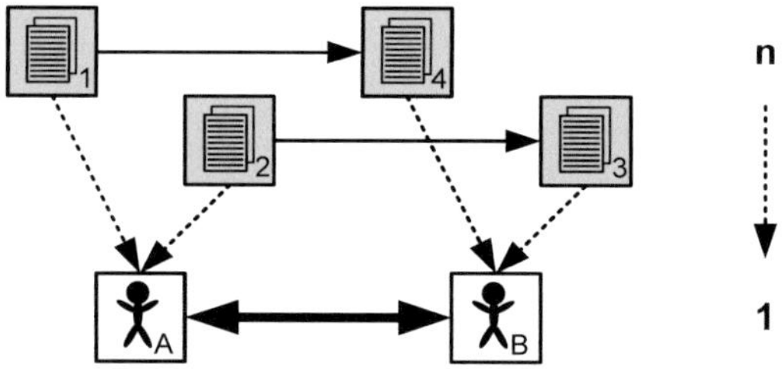

Figure 45: Aggregation of Information Flows into Communication

The last step calculates the delta between communication 'should' and communication 'as-is' by subtracting the communication 'as-is' matrix from the communication 'should' matrix.

$$Delta\text{-}DSM_{Communication} = DSM_{Communication,Should} - DSM_{Communication,As\text{-}Is}$$

5.3.3 MDM Model of Communication

The goal of the MDM model is to transfer the set-theoretical model into a more user-friendly modeling method. Square matrices, such as DSM, have been employed to document the relationships between elements of a system. DSM has several advantages for modeling design processes:

- DSMs capture feedback loops and iteration in a compact representation,
- DSMs are computable through matrix operations,
- DSMs have similar mathematical foundation as graph theory, which enables use of metrics from graph theory and SNA with DSM.

The set theoretical model is transferred into a DSM model (Steward 1981) using MDM deduction (Maurer 2007). Using MDMs, one can deduce indirect relations that connect entities of the domain in question through entities of other domains. Deduction is carried out by matrix multiplication: indirect relations in the domain in question are calculated by multiplying the DSM of the indirect domain with DMMs.

Figures 46 and 47 show examples of MDM deduction of 'should' and 'as-is' perspectives on communication based on Maurer (2007, pp.82ff.) (see figure 10, case 4). The notation 'X' inside the matrices represents a relation between the entities of the respective line and column of the matrix. In Figure 46 the DSM 'communication, should' (entities of the domain represented by capital letters) is deduced by multiplying the DSM of the indicator domain (elements represented by numbers) with the DMM and transposed DMM that connect both DSMs. In this example, the indicating domain could be, for example, a process map in which person C is responsible for completing tasks 3 and 4 (as shown in the DMMs). Since task 3 depends only on input from task 2, and task 4 depends only on input from task 3, the relations between tasks 2, 3, and 4 (in the DSM 'communication, should') can be aggregated into the indirect relation between person B and person C in the DSM 'communication, should'. While relations between indicators can be directed or undirected, relations between people in the DSMs 'communication, should' and 'communication, as-is' are by definition undirected.

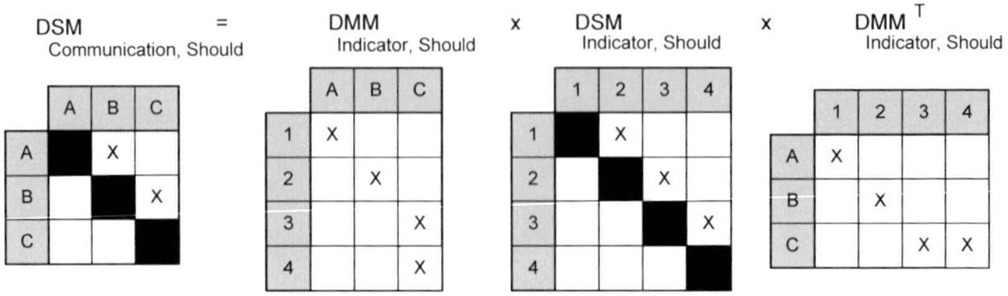

Figure 46: Calculation of DSM 'communication, should'

Figure 47 follows the same logic: entities of the DSM 'communication, as-is' are represented by lower-case letters.

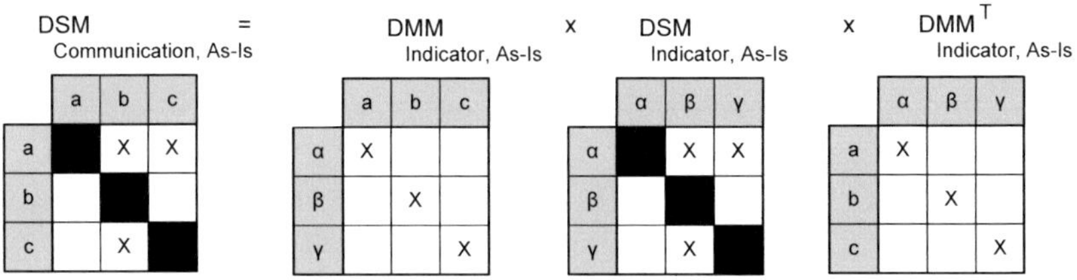

Figure 47: Calculation of DSM 'communication, as-is'

Figure 48 shows the calculation of the delta-DSM between 'should' and 'as-is' perspectives on communication. Calculation of the delta-DSM introduces the nomenclature for misalignments based on Sosa et al. (2004):

- A – Additional communication,
- M – Matching communication,
- E – Expected communication.

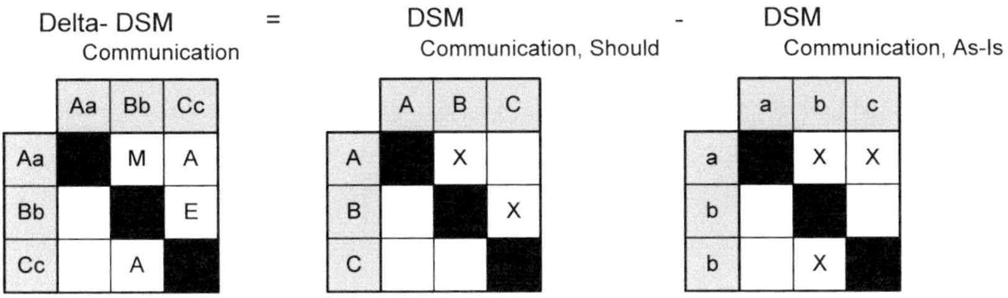

Figure 48: Calculation of Delta-DSM

Analysis of expected and additional communication can give insights into misalignments between 'should' and 'as-is' perspectives.

5.3.4 Metrics for Delta-Analysis

5.3.4.1 Network-based Metrics

Three network metrics of communication alignment can be established by summarizing and normalizing the respective relations in the delta-DSM, where n is the number of nodes in the network. Normalizing enables comparison of metrics across projects and project phases.

$$\text{Sum of Matching Communication} = \sum M/n^*(n-1)$$

$$\text{Sum of Additional Communication} = \sum A/n^*(n-1)$$

$$\text{Sum of Expected Communication} = \sum E/n^*(n-1)$$

These metrics give a general overview on whether more and/or less actual communication than planned communication took place. But these metrics do not support analysis of structural misalignments between 'should' and 'as-is' perspectives on communication.

Several network based metrics exist for the analysis of structures, and these metrics can be applied to communication networks.

Kreimeyer (2009) presents a comprehensive set of structural metrics for design processes. These metrics can be applied to compare 'should' and 'as-is' perspectives on communication. The delta between the respective values for 'should' and 'as-is' perspectives on communication, serves for analyzing structural misalignments of the two communication models.

As mentioned above, the goal of the method elaborated in this research is not to determine whether the structure of actual or planned communication is appropriate or better fitting. Instead, the goal of the method is to learn from differences between actual and planned communication structures by investigating reasons for differences. The three metrics defined above can give first directions for investigations. However, these metrics bear the risk of being used as performance indicators for actual communication. This use would imply that planned communication is correct, which may not be the case.

5.3.4.2 Network Level Application of Entity-based Metrics

Entity-based metrics mostly stem for the field of network theory. These metrics compute values for a specific entity of the network, so values of the same entity can be compared for 'should' and 'as-is' perspectives on communication. Entity-based metrics must be carefully chosen and applied, because they focus on only one entity instead of the overall communication network. However, use of entity-based metrics can be insightful when putting the values of each entity in context of the whole communication network.

Centrality metrics are useful for network analysis (see section 2.1.3.4). Entities with a high degree centrality have numerous relations with other entities. A person with a high degree centrality communicates with a large number of people. The following formula shows the mathematical calculation of degree centrality in an undirected network (based on Wasserman and Faust (1994) formulas 5.2 and 5.3), where x_{ij} = number of degrees that node i receives from node j, and n - number of nodes in the network.

$$\text{Degree Centrality } C_D\,(o)_i = \sum_{j=1}^{n} x_{ij} \Big/ (n\text{-}1)$$

$$0 \leq C_D \leq 1$$

Undirected degree centrality is appropriate to evaluate communication, because communication networks are undirected. Degree centrality serves the analysis of a people's positions within a communication network, but centralities of people must be set in context to each other for goal-oriented delta-analysis. Degree centrality can be used for delta-analysis of actual and planned communication on an entity-base, i.e., a comparison per person. However, a list of people's degree centralities focuses purely on the people's positions in the organization and higher rank on the list might be associated with higher performance or achievement. Ranking of persons is not the goal of this research, and also a ranking might get in the way of open-minded root-cause analysis.

5.4 Procedural Aspects of the Method

5.4.1 Application Procedure

The previous section presented the foundations of the method regarding modeling of communication, comparison of structures, and metrics. The presented MDM approach makes actual communication transparent by modeling it as the 'as-is' perspective on communication. This transparency serves for learning by comparing models of actual and planned communication. Learning can improve the quality of the planned process, organization architecture and integrative mechanisms.

The development of models of communication is subject to perceived complexity, since the structural model of a system is developed by a person or a group of people. Thus, perceived complexity must be considered when evaluating the quality of the model. Also, modeling and analyzing only the structural aspects of the AEC design system neglects other attributes, e.g., time and cost attributes of tasks. These attributes must be considered when deciding on changes and improvements to the system.

5.4.1.1 Requirements for Procedure

How the method is used determines its success. Thus, this section presents an application procedure for the method. The procedure considers the method's goals regarding:

(1) Quality of input data

Input data must be reliable and representative of planned and actual communication.

(2) Implementation of learning

Learning focuses on root-causes from the project system and the project environment.

5.4.1.2 Steps of Procedure

The application procedure consists of 10 steps that can be divided into four groups (figure 49):

(1) Kick-off (step 1)

During step 1 'kick-off' the project leader prepares application of the method by identifying possible data sources for models of communication and by checking accessibility and reliability of these data sources.

(2) 'Should' perspective on communication (steps 2-4)

During step 2 'choose indicators and obtain matrix (-es)' the project team picks data sources that reliably indicate 'should' communication. From these data sources the team or a responsible person builds one or several indicator matrices.

During step 3 'obtain indicator-responsibilities DMM' the project team and project leader determine which person is responsible for entities of the indicator DSM from step 2. For example, if the project team chooses a process map as indicator for 'should' communication, then the DMM captures which person is responsible for completing which task of the process map; i.e., the DMM establishes a domain-crossing relation between people and tasks.

Step 4 'deduce 'should' communication DSM' maps relations between entities of the indicator matrix to the organization domain through matrix multiplication.

(3) 'As-is' perspective on communication (steps 5-7)

Steps 5 to 7 mirror tasks 2 to 4, but focus on 'as-is' communication.

During step 5 'choose indicators and obtain matrix (-es)' the project team picks data sources that reliably indicate 'as-is' communication. From these data sources the team or a responsible person builds one or several indicator matrices.

During step 6 'obtain indicator-responsibilities DMM' the project team and project leader determine which person is responsible for entities of the indicator DSM from step 5. For example, if the project team chooses the event log of a database as indicator for 'as-is' communication, the DMM captures which people are involved in events.

(4) Learning (steps 8-10)

During step 8 'Build delta-DSM' the project leader or the person responsible for communication improvement computes the delta-DSM from DSM 'communication, should' (step 4) and DSM 'communication, as-is' (step 7). He/she also computes the metrics presented in section 5.3.4 and conducts a preliminary analysis of metrics to prepare the following workshop (step 9).

In step 9 'conduct delta-analysis with project team' the project team conducts a workshop with all team members, team leader, chief engineer, and, if applicable, leaders of other teams from the project. During the workshop the host, which should be the person who conducted the preliminary analysis of metrics, presents force-directed graphs of 'should' and 'as-is' perspectives on communication side by side. This visualization can be enhanced with project organization charts, which show the team structure of the project, a seating chart of the collocated office, or other representations of integrative mechanisms. The host of the meeting uses the method five whys (Ohno 1988, p.17) for root-cause analysis of misalignments between 'should' and 'as-is' graphs. Following root-cause analysis, the team defines actions to tackle identified root-causes.

Step 10 implements actions from step 9 and documents results of the workshop, e.g., changes in processes, organization architecture, integrative mechanisms, project documentation, and others. Documentation aids in identifying change over time and in learning about implementation of the method.

Figure 49 shows the above described 10 steps:

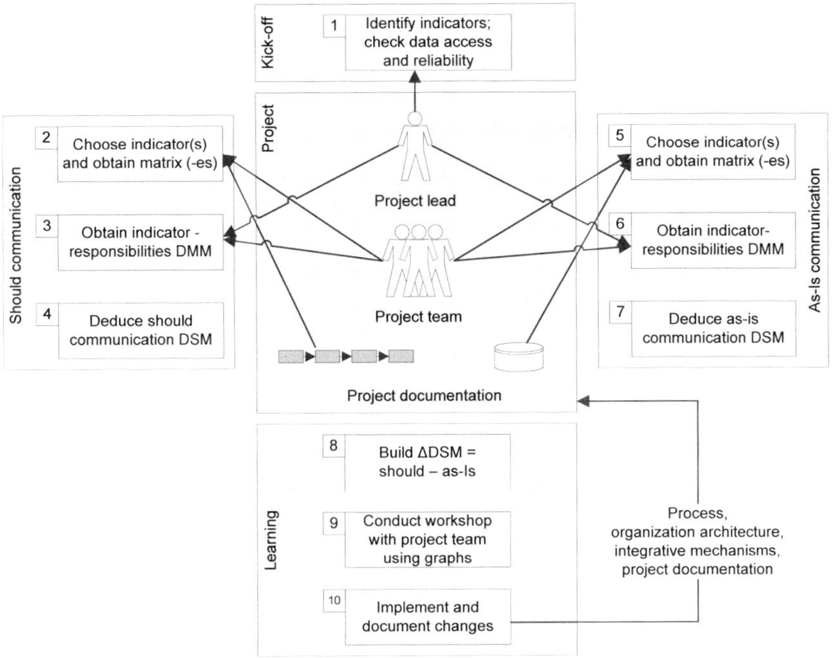

Figure 49: Application Procedure of Method for Improvement of Communication Structures using Delta-Analysis

5.4.2 Use of Indicators for Modeling of Communication

One factor which determines user friendliness and efficiency of the method is the effort for data acquisition. While models of planned information flow are usually available, e.g., through prescriptive process maps, models of actual communication are often harder to acquire. In this case, indicators can serve as proxies for communication. Business processes and design processes bear similarities, despite their differences (see section 2.2.2.1). Thus, process mining (Aalst 2005, 2011) seems also applicable for developing descriptive models of engineering design processes. Many sources for indicators exist, and the following list gives some examples:

- from the process domain: process maps, documents in circulation, or event logs from database,
- from the product Domain: modular product structure or error indications from the product model, e.g., BIM clashes,
- from the organization domain: office layouts and seating plans, email lists, organizational structure charts, or surveys which capture communication between people.

The method compares models for 'should' and 'as-is' communication. Models only represent a subset of reality. It is important for the successful application of the method that both models of communication, 'should' and 'as-is', are relevant towards the shared purpose of communication improvement.

5.4.3 Workshop Approach

Comparison of 'should' and 'as-is' perspectives on communication is a collaborative effort for the design team. Starting point of analysis is to achieve a shared understanding among members of the team about how work actually proceeded. Only when this shared understanding about the actual state of communication is achieved, then analysis of misalignments can be fruitful. Eckert and Stacey (2010) describe the importance of shared understanding:

> "An understanding of a model is a cognitive construct rather than an inherent property of the model, and a shared understanding is constructed through social processes of discussion and clarification."

Reasons for misalignments can be manifold: either people did not exchange information as they 'should' have, 'should' communication is wrongly defined, or both. Expert knowledge about the purposes of communication is necessary to identify the root-causes for misalignments. Thus, the author proposes a workshop setting in which to conduct comparison and analysis: visualization of 'should' and 'as-is' perspectives on communication with force-directed graphs aids in identifying misalignments. Root-cause analysis, e.g., using five whys (Ohno 1988, p.17), aids in finding reasons for misalignments.

5.4.4 Network Visualization using Force-directed Graphs

To fully harvest the opportunities of the method, people who execute planned communication must be involved during analysis of the misalignments between 'should' and 'as-is' perspectives on communication. Involvement and participation necessitates an understanding of both perspectives on communication, 'should' and "as-is". It is important to not only understand why things should have been done the way they were planned, but also why things happened as they did. Visualization of communication networks helps people in gaining an understanding of the structure of communication and why the structure developed the way it did.

Force-directed graphs are useful in visualizing communication, because they show the overall communication network by setting all people in context to each. Lines between people depict communication. Algorithms can incorporate degree centrality when shaping the graph. These algorithms position people with a high degree centrality in the center of the graph and people with a low degree centrality distanced from the center. Also, force-directed graphs normalize levels of degree centrality in their visualization between 'should' and 'as-is' perspective on communication, because the algorithm places people by relative degree centrality in each perspective on communication.

This visualization enables an intuitive visualization of 'who communicates with whom' and 'who is more central in the communication network'. However, a high level of communication between two people does not necessarily mean that these people communicate effectively. A high level of communication could also result from misunderstandings and additional communication for resolving misunderstandings.

5.4.5 Correspondence to Plan-Do-Check-Act Cycle

The presented method focuses on the C and A parts of the PDCA cycle. The workshop relates to the 'check' stage, and implementation of actions defined during the workshop relates to the 'act' stage of a PDCA cycle. Efficient data acquisition and a standardized method for deducing 'should' and 'as-is' perspectives on communication encourage quick learning cycles, which is what the PDCA cycle proposes.

5.5 Summary

This chapter presented the method for improvement of communication structures using delta-analysis. The method closes the research gap, because it enhances the self-reference loop of the project design organization. Specifically, this method

- achieves transparency about actual communication by using indicators for communication,
- analyzes misalignments between actual and planned communication based on the scientific experiment method by (1) using models of 'should' communication that are based on specifications for planned communication, e.g., process maps, (2) capturing actual interaction, (3) involving design team members during analysis of misalignments (4) applying root-cause analysis,
- considers the project as an open system by using root-cause analysis and team involvement during analysis of misalignments.

The next chapter of this dissertation presents two case studies (B1 and B2) that apply the method for improvement of communication structures using delta-analysis.

6 Case Studies B1 and B2[28]

6.1 Case Study B1 – VNGC Project BIM Development Process

6.1.1 Case Study Description

The setting of this case study is the US-$1.7 billion Van Ness and Geary Campus (VNGC) Project in San Francisco (California, USA), formerly known as Cathedral Hill Hospital (CHH) Project. In part due to seismic code regulations, the design of hospitals in California is complex. The project applies an IPD-type contract, the Integrated Form of Agreement (IFOA) (Lichtig 2006). The IFOA sets incentives for collaboration between project participants through pain-and-gain sharing, joint management of financial risk, joint management of disputes, and other mechanisms (Lichtig 2005). Being an IFOA project, VNGC also applies lean construction principles to operating system and organization.

At the time of the case study - April to May of 2011 - the project was in the detailing phase of design. In the detailing phase, designers created an integrated 3D-model of the building using BIM. BIM developers of different trade partners were collocated in one office with other experts so that they could communicate easily and solve conflicts quickly. A challenge in AEC design and especially hospital design is to fit interdependent systems into small spaces, while meeting numerous functional requirements yet maximizing open spaces (rooms) for operational building use. Interdependency of systems refers not only to connectiveness and spaces, but also to other properties and capabilities, which impact performance. Design of these dense spaces can be critical for

[28] Parts of this section have been published in Hickethier et al. (2011b, 2012).

project success. A critical question of the detailed design phase is: how will the model be built? The process of developing the BIM model needs to be designed according to the characteristics of the project and the capabilities of those involved.

BIM developers of VNGC have identified 'system flexibility' as a key determinant of their modeling sequence. The least flexible systems (more physically rigid) shall be modeled first, and systems modeled subsequently shall adapt to the space constraints thus imposed (in other words, they will 'wrap around what is already in place'). However not all components in a systems are equally (in-) flexible, so BIM developers must adapt their process to the needs of the actual modeling task. They use Plan-Do-Check-Act (PDCA) to improve their BIM detailing process and they work in cycles, each cycle comprising the detailing of one floor of the building. In the 'check' phase of the PDCA cycle, the team uses dashboards to visualize commitment reliability and the team uses plus-delta reviews to identify opportunities for improvement. In the act phase of the PDCA cycle, the team uses A3-Reports, to document and analyze challenges and alternatives, and in structuring, evaluating, and implementing solutions (Chandler et al. 2011).

The author of this dissertation collected data for the case study B1 during a two-month period. He was collocated with the detailing team in the project office and had access to file servers and people. Data collection proceeded through access to files, e.g., clash reports and standard processes, and interviews with modelers. Collected data served building models of actual and planned communication. These models were presented to the detailing team in a workshop, which was hosted by the author.

The software LOOMEO[29] was used to complete part of the analysis and for visualization of communication models. LOOMEO was chosen,

[29] More information available at www.loomeo.com.

because it includes several capabilities which were needed during this case study:

- LOOMEO serves the analysis of complex systems through data acquisition, analysis based on DSM, MDM, Network Theory, and Graph Theory techniques, and visualization.
- LOOMEO includes the capability of deducing DSM based on the MDM method (see section 2.1.3.3).

Microsoft Excel was used to compute the Delta-DSM. The software Gephi (Bastian et al. 2009) was used to analyze the model of actual communication with weighted dependencies.

6.1.2 Modeling of Communication

BIM developers aim to achieve an error-free model during design in order to avoid costly rework during construction. As a part of this, they perform clash detection (Eastman et al. 2008, p. 216). That is, they use BIM to identify spatially conflicting building parts. 'Hard clashes' refer to parts occupying the same space that would collide during construction and therefore cannot be built as designed. 'Soft clashes' refer to parts being within a certain range of each other, and this range can be set, e.g., to building code requirements: For example, in California no part of a building may be closer than five centimeters to the structural steel in order to not damage the fireproofing that coats the structural steel.

Lean practitioners will want to avoid errors (including clashes) upfront, while developing the BIM model (Tommelein and Gholami 2012). Interaction that should be avoided may be criticized for being little meaningful as indicator for communication. However, current industry practice is far from clash-free processes. Instead clashes are a standard phenomenon. Therefore, clash resolution is a type of communication worth studying.

Once a clash is identified, BIM developers must rework the contents of the model. Rather than reworking a clash, BIM developers should avoid this kind of wasteful rework. Clash avoidance needs a well-defined development process according to which to populate the BIM model. Specifically, the development process must (1) be designed to the characteristics of the actual project and people involved, and yet (2) allow flexibility for exceptions from the standard rules.

Regarding (1), BIM developers may follow the PDCA cycle (e.g., Deming 1982, p.88) to continuously improve their BIM development process, thereby adapting it to the characteristics of the actual project as it unfolds through learning loops. Regarding (2), a process should allow for flexibility in case the proposed development sequence proves impractical. BIM developers from different trade partners often find solutions for clashes based on who can move their systems most easily while keeping system performance. The identified solution can require deviating from the process as specified.

Use of the PDCA cycle requires a 'check' of the development process in use. Here, the author focuses on communication pertaining specifically to BIM modeling meaning 'drawing of BIM components,' rather than on the tasks defining how to organize the model or how to go about modeling. A comparison between planned communication ('should' perspective) and the actually happening communication (real communication as it is taking place during the design process) ('as-is' perspective) can test alignment between planning and reality. Differences between the perspectives can be used as a starting point for a 'check' of the planned process and then be followed by 'act'-ing to improve the process.

BIM developers must identify misalignments between the 'should' and 'as-is' perspectives, and then find root-causes for them, in order to improve their processes. Documentation of real communication is time consuming and impractical, if not infeasible. However, the identification of clashes in the BIM can be used as an indicator for communication between developers, because the resolution of each conflict will need communication between the developers who worked on the conflicting components.

This case study B1 applies the method for communication improvement using delta-analysis. This method compares models of actual and planned communication in order to find and analyze misalignments between these models. In this case study, the method consists of two models:

(1) Model for actual communication: BIM clashes serve as indicator
(2) Model for planned communication: BIM development process serves as indicator

Figure 50 shows an example of combining the two models for finding misalignments between 'should' and 'as-is' perspectives on communication.

BIM developer A, who develops system 1, and BIM developer B, who develops system 2, are indirectly connected to each other when systems 1 and 2 clash with each other in the BIM. In this case, developers A and B need to communicate with each other to resolve the conflict (figure 50, 'as-is' case). Also, the BIM development process connects the developers indirectly: when developer A works on task 1 and developer B needs task 1 to be completed in order to begin to work on task 2, then developer B depends on developer A's information (figure 50, 'should' case).

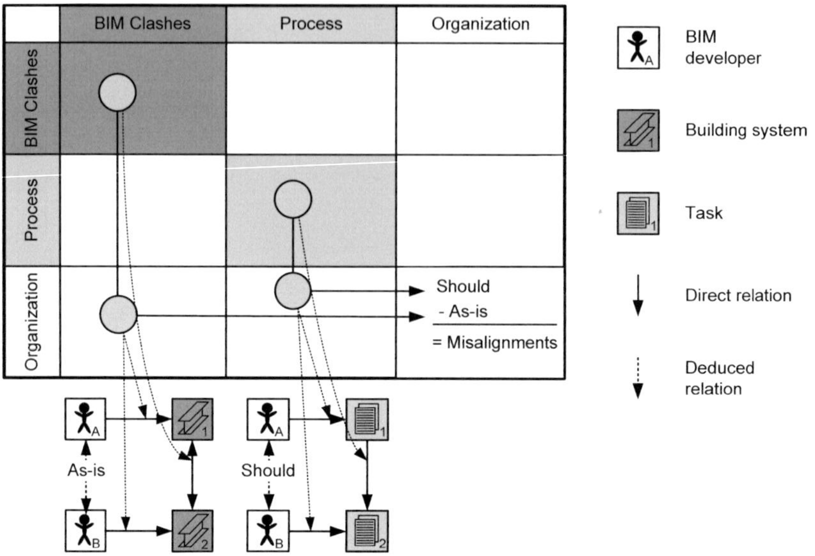

Figure 50: Model of Communication in Case Study B1

The Multiple Domain Matrix (MDM) can integrate models of actual and planned communication and then analyze relations between entities across different domains. Entities and relations between entities in any given domain are represented by a DSM (Steward 1981). Domain Mapping Matrices (DMM) (Danilovic and Browning 2004) then connect the DSMs. Together these matrices form the Multiple Domain Matrix (Maurer 2007).

Use of deduction logic (Maurer 2007, p.82) yields two DSMs for the organization domain: (1) the DSM 'communication, should' results from indirect relations through the Process domain and (2) the DSM 'communication, as-is' results from indirect relations through the BIM clashes domain. Comparison of these two DSMs may show misalignments between actual and planned communication.

6.1.3 Practical Implementation

Practical Implementation followed the 10 Step approach outlined in section 5.4.1.2. The software LOOMEO was used to deduce relations and visualize and analyze graphs.

(1) Kick-off

The author of this dissertation initiated the improvement project by presenting the method for improvement to the leader of the MEP cluster group of the VNGC project. Next, he identified possible data sources for modeling communication among members of the MEP cluster group. The author identified the BIM development process as a data source for modeling the 'should' perspective on communication. Emails sent between members of the cluster group or BIM conflicts between building systems were identified for modeling the 'as-is' perspective on information flow.

(2) Choose indicators and obtain matrix (-es) for 'should' perspective

The author, jointly with the leader of the MEP cluster group, chose the BIM development process as indicator for the 'should' perspective on communication. Cluster group members used a LAP approach for implementing the planned process. Cluster group members requested work, made commitments for work, and checked fulfillment of commitments in a weekly group meeting. Process maps served as dashboards for tracking commitments. During group meetings task status was indicated on the print-outs of process maps using markers.

Figure 51 shows the process map of the BIM development process. The DSM 'indicator, should' was built from the modeling tasks indicated in the process map. Relations in the process map focus on the coordination cycle between three batches of tasks. These three batches establish a modeling sequence of building systems. On the task level, the process map does not follow an established process mapping notation, but the coordination cycle indicates that iteration is planned to be part of the modeling process. This iteration shall be value-adding positive iteration (Ballard 2000b) which improves design process, facility, and construction performance.

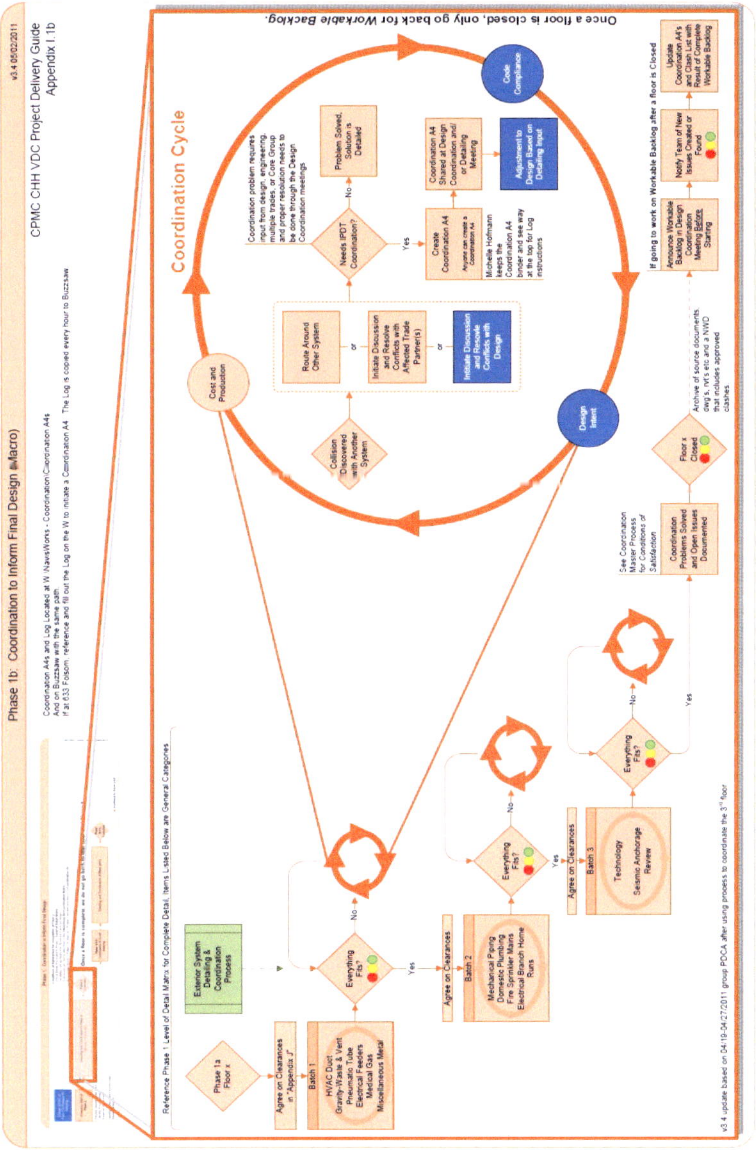

Figure 51: Standard BIM Development Process from VNGC Project Delivery Guide (Sparapani 2011, p.30)

Transfer of the process map into a task-based DSM demanded additional information from the cluster leader and the author of the process map in order to determine the level of planned iteration. In a first step, all tasks from the process map which entail working in the BIM were listed in a task-based DSM. All tasks that do not entail working in the BIM were omitted, e.g., the task "agree on clearances" was omitted from the task-based DSM. In a second step, five scenarios of iteration between the tasks of the DSM were compared. Tables 22 and 23 show the five scenarios of the 'should' perspective on communication in DSMs 'indicator, should' and in force-directed graphs. Discussion with the cluster leader and the author of the process map revealed that value-adding iteration was planned to occur mostly within the three batches of tasks. Within each stage, developers shall still follow the modeling sequence, which should lead to a moderate amount of iteration. Iteration can also occur between batches, but then long feedback cycles can affect several systems and cause excessive rework.

Scenario 2 'not overlapping feedback loops' was chosen as the DSM-model closest to the process map, because it models iteration within stages but not across stages, and the level of iteration within stages is moderate. Figure 52 shows a generic model of the planned modeling process consisting of three sequential stages with three tasks within each stage.

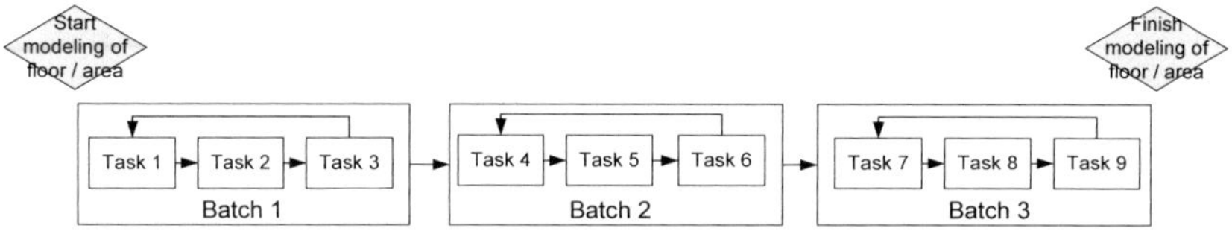

Figure 52: Flowchart of planned Modeling Process

Tables 22 and 23 illustrate the five scenarios of iteration between tasks. Visualizations of the five scenarios include DSMs 'indicator, should' and force-directed graphs of the DSMs 'communication, should'. Increased iteration between tasks in DSM 'indicator, should' (visible through lower diagonal marks in the DSMs) causes increased connectedness of BIM developers in DSM 'communication, should' (visible in force-directed graphs).

Table 22: Analysis of Iteration in DSM 'indicator, should' and Effects on Communication (part 1)

Scenario	no iteration	not overlapping feedback loops	full iteration, not overlapping feedback loops
DSM 'indicator, should'	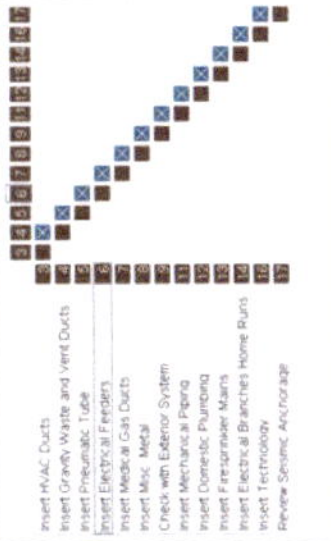	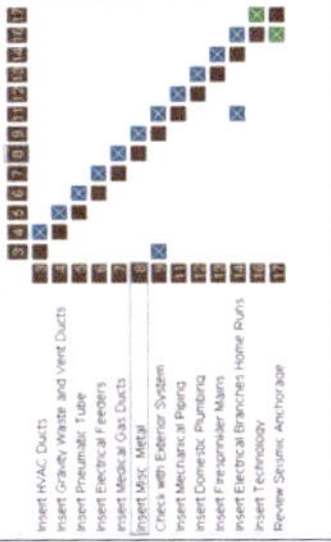	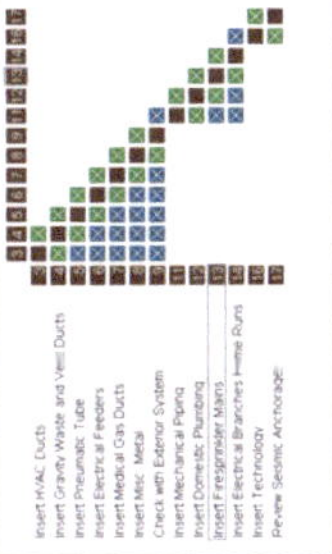
Force-directed graph based on DSM 'communication, should'		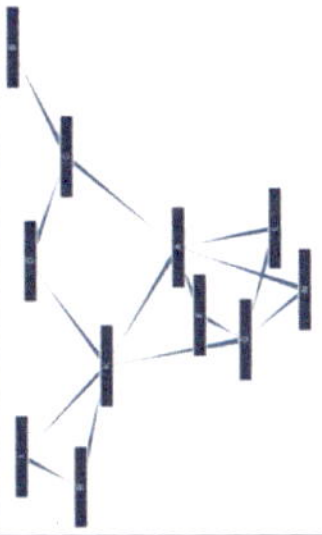	

Table 23: Analysis of Iteration in DSM 'indicator, should' and Effects on Communication (part 2)

Scenario	overlapping feedback loops	full iteration, overlapping feedback loops
DSM 'indicator, should'	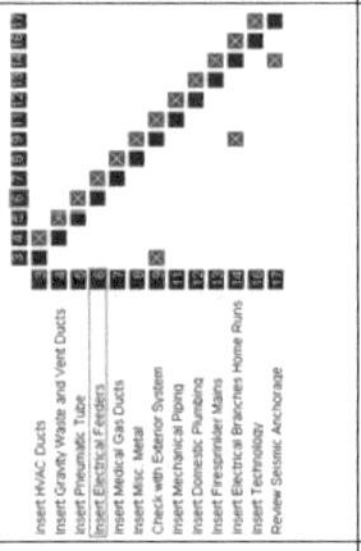	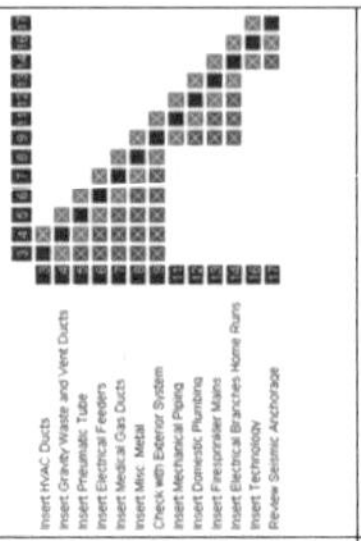
Force-directed graph based on DSM 'communication, should'	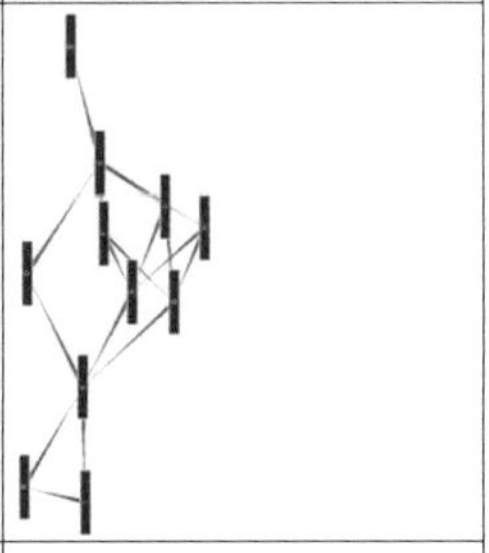	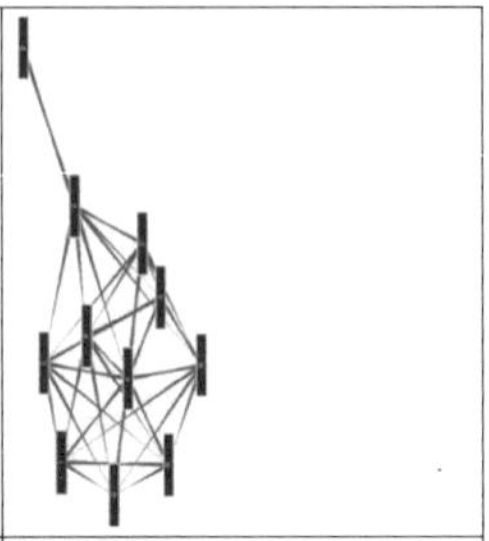

(3) Obtain indicator-responsibilities DMM for 'should' perspective

The DMM denotes each BIM developer's responsibility for tasks of the BIM development process. Responsibilities were collected and verified through interviews with BIM developers.

BIM developers of the same system sometimes divided work by floor, e.g., BIM developer 1 works on floors with even numbers and BIM developer 2 on floors with uneven numbers. The analysis focuses on modeling of one floor, and therefore some BIM developers do not have tasks assigned in the DMM. Mapping of BIM developers to tasks is mostly 1 to 1, i.e., one BIM developer models only one building system. But one BIM developer of the cluster group, BIM developer K, models three building systems. Thus, he is responsible for three tasks which stem from two different batches of the BIM development process. This characteristic of assigning BIM developers to tasks will be discussed below.

(4) Deduce 'DSM communication, should'

Multiplication of DMM 'indicator, should' with DSM 'indicator, should' of scenario 2 and transposed DMM 'indicator, should' yielded DSM 'communication, should'. Figure 53 shows DSM 'communication, should'; letters (except 'X') represent BIM developers.

	A	B	C	D	E	F	G	H	I	J	K	L	M	N	O	P	Q	R	S
A			X			X					X	X		X					
B			X																
C	X	X		X															
D			X								X								
E																			
F	X																X		
G																			
H																			
I											X							X	
J																			
K	X			X					X										
L	X																X	X	
M																			
N	X																X		
O																			
P																			
Q						X					X			X					
R									X		X								
S																			

Figure 53: DSM 'communication, should'; Letters represent BIM Developers

(5) Choose indicators and obtain matrix (-es) for 'as-is' perspective

The author jointly with the MEP cluster group leader chose BIM clashes between building systems as indicator for the 'as-is' perspective on communication. BIM clashes only cover a subset of communication between BIM developers, but communication regarding clashes is important as it represents wasteful rework. The purpose of the BIM development process (figure 51) is to avoid BIM clashes, so communication regarding clashes fits as indicator for actual communication.

Figure 54 shows a clash report from April 27, 2011. This clash report summarizes clashes for one floor of the building and was used as the basis for modeling the 'as-is' perspective on communication.

Clash Batch Test Matrix

	Steel	Critical Studs Head of Wall	Kickers Soffits	Ceilings	Ceiling Compression Posts	HVAC	Fire Sprinkler Mains	Waste & Vent	Fuel Oil	Electrical Feeder	Electrical Equipment	Pneumatic Tube	Mechanical Piping	Domestic Water	Electrical Branch Home Runs	Medical Gas	Technology	Fire Sprinkler Branch	Lighting	Temporary Power	Exterior Framing	Curtain Wall	Exterior Stone	Metal Panel
	SXX	MFCF	MFFK	AIN	ACCL	MCSM	FP-M	PCVW	FOP	ECFC	ECEQ	PT	MCMP	PCDV	ECBC	PCMG	TC	FP-B	EDEL	ECTP	MFFE	CWEX	SPEX	MPEX
Steel		X	x	x	x	0	5	2	0	6	0	0	5	0	2	0	0	29	4	0	354	9	30	71
Critical Studs			X	X	x	41	10	40	0	6	2	4	14	10	2	3	10	1	8	0	x	311	39	94
Kickers & Soffits				X	x	607	7	50	1	28	0	2	45	20	0	21	65	14	410	0	x	8	0	0
Ceilings					X	x	x	x	x	x	x	x	x	x	x	x	x	x	x	x	x	x	x	x
Ceiling Compression Posts						x	x	x	x	x	x	x	x	x	x	x	x	x	x	x	x	x	x	x
HVAC							4	0	0	0	0	3	12	7	2	2	4	1	12	0	5	0	13	40
Fire Sprinkler Mains								3	0	0	0	0	3	4	2	0	1	x	0	0	0	0	0	0
Waste & Vent									0	0	0	1	2	5	0	1	0	2	3	0	8	0	3	0
Fuel Oil										0	0	0	0	0	0	0	0	0	0	0	2	0	12	0
Electrical Feeder											0	0	1	0	0	0	1	0	0	6	0	0	0	0
Electrical Equipment												0	0	0	0	0	0	0	0	0	0	0	0	0
Pneumatic Tube													0	0	0	6	0	0	0	0	0	0	0	0
Mechanical Piping														11	0	5	0	0	4	4	1	0	0	0
Domestic Water															0	0	0	0	0	0	0	0	0	0
Electrical Branch Home Runs																1	0	0	0	0	1	0	0	0
Medical Gas																	0	0	0	0	0	0	0	0
Technology																		0	4	29	0	0	0	0
Fire Sprinkler Branch																			0	1	0	0	0	1
Lighting																				0	2	25	0	0
Temporary Power																					0	0	0	0
Exterior Framing																						31	143	484
Curtain Wall																							11	43
Exterior Stone																								15
Total																								3.377

Up | Down | New | Same | X Test not required

Figure 54: Clash Test Batch Matrix from 2011-04-27 (courtesy of Michelle Hofman, VNGC)

Figure 55 shows the DSM 'indicator, as-is' based on the clash report (figure 54). This model of clashes using DSM has limitations, because the DSM denotes binary relations. Accordingly, the DSM contains information regarding whether or not clashes exist between two systems, but it does not contain information regarding the number of clashes between systems nor does it contain information on the severity of clashes. Author and cluster leader jointly decided to represent all clashes in the DSM. Therefore, the threshold for denoting an 'X' mark in the DSM is one clash between systems.

		1	2	3	4	5	6	7	8	9	10	11	12	13	14	15	16	17	18	19	20	21	22	23	24
Steel	1							X	X		X			X		X			X	X		X	X	X	X
Critical Studs, Heads of Wall	2						X	X	X		X	X	X	X	X	X	X	X	X	X			X	X	X
Kickers & Soffits	3						X	X	X	X	X		X	X	X		X	X	X	X			X		
Ceilings	4																								
Ceiling Compression Posts	5																								
HVAC	6		X	X				X					X	X	X	X	X	X	X	X		X		X	X
Fire Sprinkler Mains	7	X	X	X			X		X					X	X	X		X							
Waste & Vent	8	X	X	X				X					X	X	X		X		X	X		X		X	
Fuel Oil	9			X																		X		X	
Electrical Feeder	10	X	X	X										X				X			X				
Electrical Equipment	11		X																						
Pneumatic Tube	12		X	X			X		X								X								
Mechanical Piping	13	X	X	X			X	X	X		X				X		X			X	X	X			
Domestic Water	14		X	X			X	X	X					X											
Electrical Branch Home Runs	15	X	X				X	X									X					X			
Medical Gas	16		X	X			X		X				X	X		X									
Technology	17		X	X			X	X			X									X	X				
Fire Sprinkler Branch	18	X	X	X			X		X												X				X
Lighting	19	X	X	X			X		X					X				X				X	X		
Temporary Power	20										X			X				X	X						
Exterior Framing	21	X					X		X	X				X		X				X			X	X	X
Curtain Wall	22	X	X	X																X		X		X	X
Exterior Stone	23	X	X				X		X	X												X	X		X
Metal Panel	24	X	X				X												X			X	X	X	

Figure 55: DSM 'indicator, as-is'

(6) Obtain indicator responsibilities DMM for 'as-is' perspective

The DMM denotes BIM developers' responsibilities for building systems. Responsibilities were collected and verified through interviews with the BIM developers.

In-line with DMM 'indicator, should', BIM developer K is responsible for three building systems of the clash report. The three systems align with the three modeling tasks from DSM 'indicator, should'.

(7) Deduce DSM 'communication, as-is'

Multiplication of DMM 'indicator, as-is' with DSM 'indicator, as-is' and transposed DMM 'indicator, as-is' yielded DSM 'communication, as-is'. Figure 56 shows DSM 'indicator, as-is'; letters except 'X' represent BIM developers.

	A	B	C	D	E	F	G	H	I	J	K	L	M	N	O	P	Q	R	S
A				X	X					X	X		X		X	X	X	X	
B				X						X			X		X		X	X	
C				X	X					X	X				X	X	X		
D	X	X	X		X	X				X	X				X	X	X	X	
E	X		X	X		X				X	X	X	X	X				X	
F				X	X					X		X		X	X	X	X		
G																			
H																			
I										X	X				X		X		
J	X	X	X	X	X	X			X		X	X	X	X		X	X	X	X
K	X		X	X	X				X	X		X	X		X	X	X		
L					X	X				X	X			X	X	X	X		X
M	X	X			X					X	X			X	X	X	X		
N					X	X				X		X	X		X	X			
O	X	X	X	X		X			X		X	X	X	X			X	X	X
P	X		X	X		X				X	X	X	X	X				X	
Q	X	X	X	X		X			X	X	X	X	X		X				
R	X	X		X	X					X					X	X			
S										X		X			X				

Figure 56: DSM 'communication, as-is'; Letters represent BIM Developers

(8) Build Delta-DSM

The delta-DSM was computed by subtracting the DSM 'communication, as-is' from the DSM 'communication, should'. Figure 57 shows the resulting delta-DSM with matching communication (M), additional communication (A), and expected communication (E). Colors do not imply any evaluation, but are only added to ease the identification of possibly existing patterns.

	A	B	C	D	E	F	G	H	I	J	K	L	M	N	O	P	Q	R	S
A			E	A	A	E				A	M	E	A	E	A	A	A	A	
B			E	A						A			A		A		A	A	
C	E	E		M	A					A	A				A	A	A		
D	A	A	M		A	A				A	M				A	A	A	A	
E	A		A	A		A				A	A	A	A	A				A	
F	E			A	A					A		A		A	A	A	M		
G																			
H																			
I										A	M				A		A	E	
J	A	A	A	A	A	A			A		A	A	A	A		A	A	A	A
K	M		A	M	A				M	A		A	A		A	A	A		
L	E				A	A				A	A			A	A	A	M	E	A
M	A	A			A					A	A			A	A	A	A		
N	E				A	A				A		A	A		A	A	E		
O	A	A	A	A		A			A		A	A	A	A			A	A	A
P	A		A	A		A				A	A	A	A	A				A	
Q	A	A	A	A		M			A	A	M	A	A	E	A				
R	A	A		A	A				E	A	E				A	A			
S										A		A			A				

Figure 57: Delta-DSM

The delta-DSM shows a large number of additional marks, and the number of expected marks is higher than the number of matching marks. The metrics reflect these misalignments between 'should' and 'as-is' perspectives:

Sum of Matching Communication $= \sum M/n^*(n-1) = 12 / 19 * (19-1) = 0.04$

Sum of Additional Communication $= \sum A/n^*(n-1) = 138 / 19 * (19-1) = 0.4$

Sum of Expected Communication $= \sum E/n^*(n-1) = 16 / 19 * (19-1) = 0.05$

Table 24 shows the degree centralities of BIM developers. Degree centralities vary between 'should' and 'as-is' perspectives, with the largest differences for BIM developers J and O.

Table 24: Degree Centralities of BIM Developers in 'should' and 'as-is' Perspectives

Person	**Degree Centrality**		
	'Should' perspective	**'As-is' perspective**	**Delta**
A	5	9	-4
B	1	6	-5
C	3	7	-4
D	2	11	-9
E	0	10	-10
F	2	8	-6
G	0	0	0
H	0	0	0
I	2	4	-2
J	0	15	-15
K	5	11	-6
L	2	9	-7
M	0	9	-9
N	2	7	-5
O	0	13	-13
P	0	10	-10
Q	4	11	-7
R	2	7	-5
S	0	3	-3

Preliminary analysis led to the assumption that work iterates between batches of the BIM modeling process. Instead of working sequentially, BIM developers work on batches concurrently. Figure 58 shows a flowchart of the assumed modeling process.

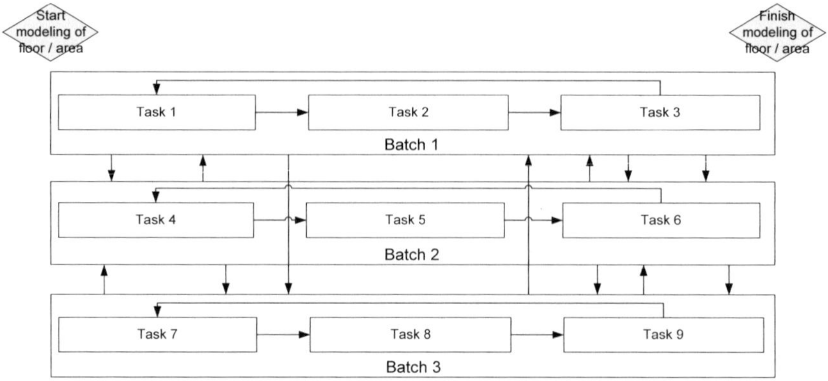

Figure 58: Assumed actual Modeling Process after preliminary Analysis

(9) Conduct delta-analysis with project team

After preliminary analysis, a date for the workshop was reserved and the cluster group, the chief engineer, and the leaders of the other cluster groups were invited. The chief engineer and cluster group leaders coordinate between cluster groups and their knowledge is needed during root-cause analysis, because problems can originate in or concern other cluster groups.

The MEP cluster group, chief engineer, and cluster group leader of the exterior cluster attended the meeting. The meeting started with a short introduction into DSM modeling and MDM deduction, followed by a presentation of data sources used. Assumptions developed during preliminary analysis (step 8) were intentionally not mentioned. Presentation of the preliminary analysis may have influenced results of the workshop, and it was the intention of the author that BIM developers conduct the analysis. Next, the author presented 'should' and 'as-is' perspectives on communication as force-directed graphs (figures 59 and 60). Graphs were visualized using the LOOMEO software. LOOMEO's drawing algorithm places entities with a high degree centrality closer to the center of the graph, but the layout of the graph does not present degree centrality in a precise manner Nevertheless, the major structural differences between 'should' and 'as-is' perspectives are visualized (figures 59 and 60).

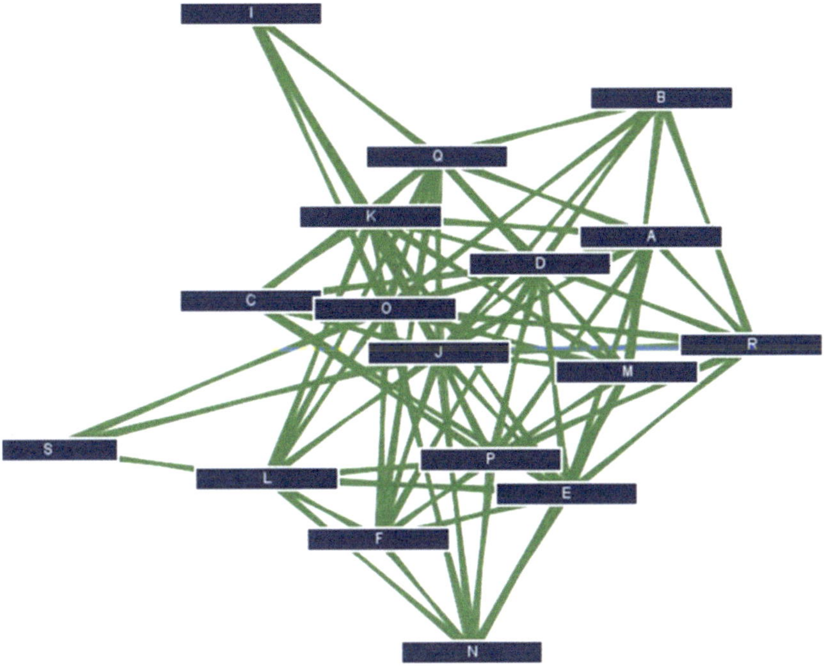

Figure 59: 'As-is' Perspective on Communication

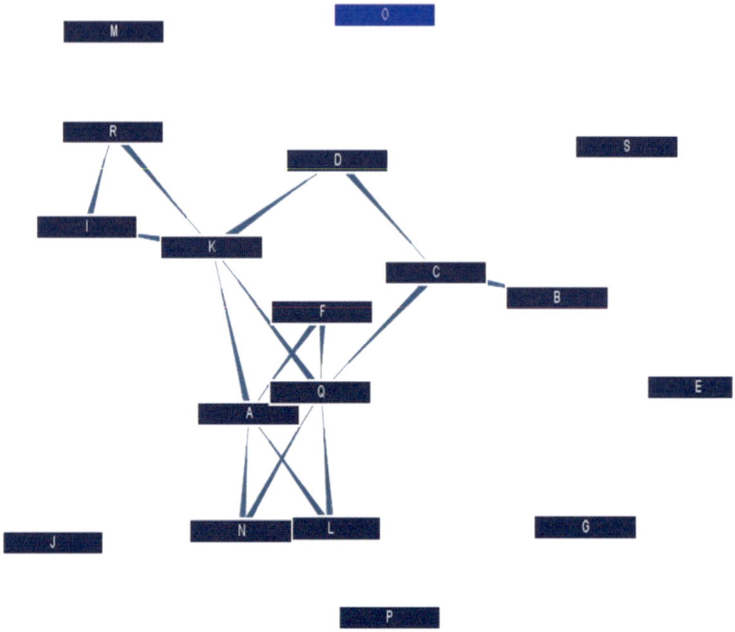

Figure 60: 'Should' Perspective on Communication

Graphs were presented along with the question: "why are there differences between the two perspectives?". Presentation of the force-directed graphs spurred an intense discussion among participants in the meeting. In an open atmosphere participants discussed several reasons for misalignments between 'should' and 'as-is' perspectives.

The assessment of the team was that BIM developers do not look at the partitioning layer of the BIM when modeling their systems. BIM can consist of several layers and BIM developers can choose which layers they want to see while working on the model. BIM developers develop the partitioning layer before the here-presented portion of the BIM development process starts. Hence, the partitioning layer is an input of the planned BIM development process. A large number of clashes between the partitioning system and other systems exist, which led to the conclusion that BIM developers seemed to not look at the partitioning layer.

These problems were more deeply analyzed with root-cause analysis using the method five whys (Ohno 1988, p.17). After root-cause analysis, the team defined actions to solve the identified problems. Table 25 presents identified problems and related actions; problem identification relates to the 'check' part of the PDCA cycle, and Action relates to the 'act' part of the PDCA cycle (Deming 2000).

Table 25: Identified Problems and related Actions of Case Study B1

Identified Problem	**Action**
BIM developers O's and J's task, which was modeling of partitioning, was not part of the detailing portion of BIM development process.	Team agreed to change the modeling process.
Other BIM developers did not load the partitioning layer into their modeling programs, because loading time for this layer is especially long.	A3 report to investigate reasons for long loading times.
BIM developer J was seated about 15 m away from others on the detailing team. BIM developer O does not work in the 'big room,' but in an office several hundred kilometers away.	Define standard process to integrate BIM developer O with the rest of the team. Also, team members introduced actions to improve communication with BIM developer J.
The use of BIM for partitioning is a relatively recent development in the industry and other trades on the project were not used to integrate their work with that of partitioning BIMs.	Raise awareness to stimulate change.

(10) Implement and document changes

Following the workshop the team implemented the actions, which were agreed upon during the workshop.

6.1.4 Results of Workshop

Presentation of the force-directed graphs made the communication pattern between team members transparent. During discussion different views on the process surfaced and the presented graphs facilitated a discussion about reasons for different views. The discussion resulted in collaboratively defined actions. Hence, visualization of communication patterns with force-directed graphs helped in aligning BIM developers' divergent perspectives, i.e., graphs helped reduce BIM developers' divergent perceptions of reality.

Integrative mechanisms played a large role during discussion of misalignments. During the discussion BIM developers' knowledge about characteristics of integrative mechanisms was helpful for successful root-cause analysis; expert knowledge from the design shopfloor helped in finding root-causes. Figure 61 to 64 show communication from 'should' perspective (figure 61), 'as-is' perspective (figure 62), organization architecture (figure 63), and seating chart (figure 64). Comparison of these four perspectives makes structural differences visible:

- BIM developer O is in the center of the 'as-is' perspective, but not connected to other BIM developers in the 'should' perspective, and not located on the seating chart (as he does not work from the collocated office),
- BIM developer J is in the center of the 'as-is' perspective, but not connected to other BIM developers in the 'should' perspective, and on the seating chart located in between cluster groups MEP (green) and exterior (yellow),
- BIM developer M is in the center of the 'as-is' perspective, but not connected to other BIM developers in the 'should' perspective. He is seated with the MEP cluster group (seating chart) but member of the design cluster group (organization architecture).

These structural differences were not presented during the workshop. However, BIM developers O's and J's positions in the structures of these integrative mechanisms surfaced during group discussion based on the knowledge of the participants.

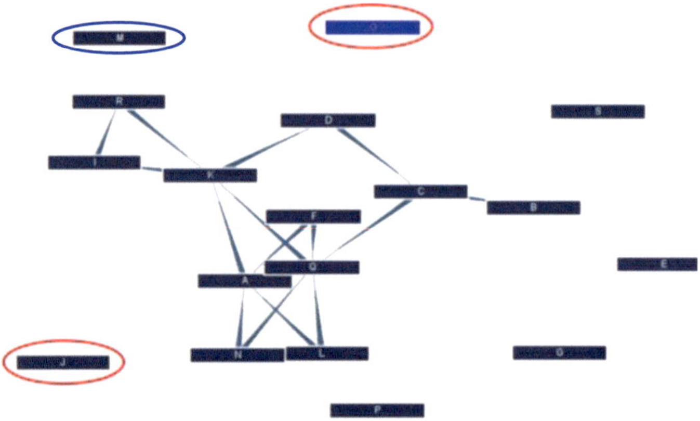

Figure 61: 'Should' Perspective on Communication (annotated)

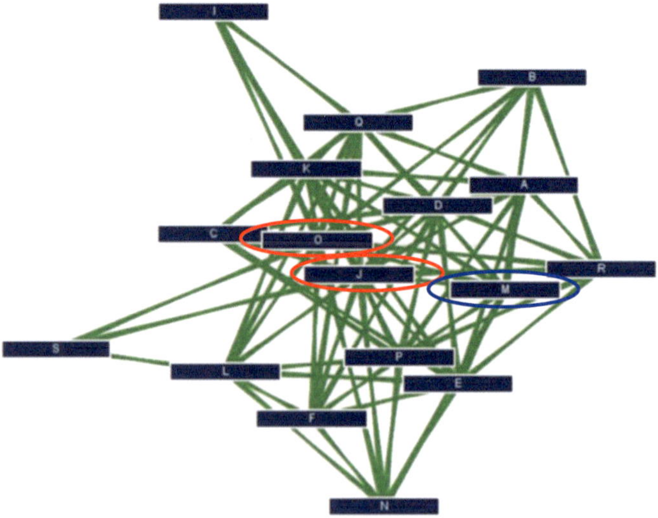

Figure 62: 'As-is' Perspective on Communication (annotated)

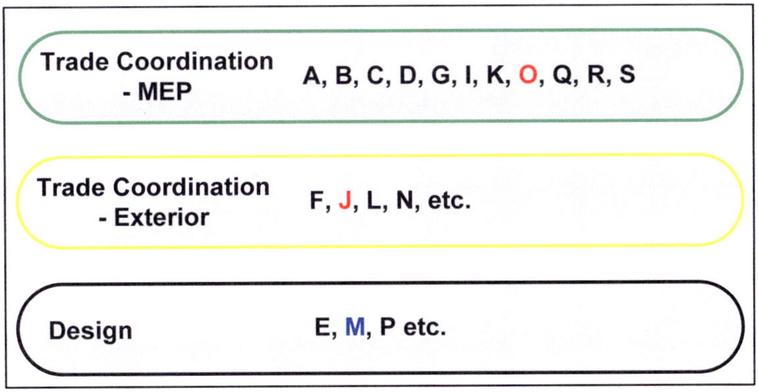

Figure 63: Cluster Group Memberships

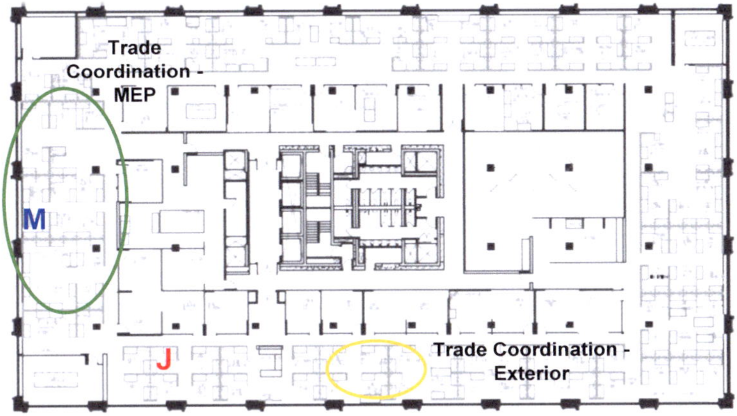

Figure 64: Seating-Chart of the Collocated Office

The identified problems show the interdependence of actual communication, planned process, organization architecture, and integrative mechanisms. Table 26 lists problems, the domain in which the identified problems originated, and the related part of lateral relations.

Table 26: Identified Problems of Case Study B1 in Context of System Model and Lateral Relations

Identified Problem	Root Domain	Part of lateral relations
Missing tasks in portion of prescriptive process model	Process	Prescriptive Process
Long loading times of BIM model	Organization - Technology	Integrative Mechanisms: Improved information and communication technologies – BIM Database
Missing integration of partitioning BIM developers	Organization	Integrative Mechanisms: Collocation
Missing experience regarding integration of partitioning contractors into BIM development process	Project Environment	n/a

6.1.5 Critical Review of Modeling

Models are representations of an original entity, but models often only cover subset of the attributes of the original entity. The choice of modeled attributes shall align with the purpose of the model (Stachowiak 1973, pp.131ff.). Successful analysis of misalignments demands models of actual and planned communication that align with the purpose of improvement of communication structures. Becker et al. (1995) describe six modeling guidelines to ensure quality of a model. Table 27 assesses models of actual and planned communication of this case study B1.

Table 27: Assessment of Modeling Guidelines

Modeling guidelines (Becker et al. 1995)	**Model of actual communication**	**Model of planned communication**
System correctness: the model is syntactically and semantically correct.	The model is based on the meta-model and the DSM modeling technique; it describes coordination requirements between building systems.	The model is based on the meta-model and the DSM modeling technique; it describes an order of modeling tasks. Five scenarios of iteration between modeling tasks were compared and the most appropriate chosen.
Relevance: only the parts of interest of the original entity are mapped to the model.	Coordination requirements are modeled in a binary attribute (yes/no). This model is a simplification as it neglects number of clashes between building systems and the severity of each clash.	The model contains only information regarding tasks and dependencies between tasks. The model does not contain information regarding task duration, resources needed, execution of dependencies (push/pull), or other attributes of tasks or dependencies. (Table continued on next page)

Economic efficiency: there is a trade-off between the effort for developing the model and making it as complete as possible.	A report regarding clashes between building systems was available. Future modeling techniques could include the number of clashes between building systems and improve visualization (see below).	The process map was available. Process modeling needed analysis regarding the amount of planned iteration.
Clarity: a user is able to understand the model.	Force-directed graph provides a more intuitive understanding of communication structures than matrices, e.g., DSM.	Force-directed graph provides a more intuitive understanding of communication structures than matrices, e.g., DSM.
Comparability: the guidelines in a modeling project are consistently utilized, e.g., naming conventions.	MDM modeling provides guidelines for deduction which imposes clear relations of entities across domains.	MDM modeling provides guidelines for deduction which imposes clear relations of entities across domains.
Systematic design: different views on the original entity are clearly distinguished.	Several sources for deducing actual communication are available and these sources were clearly separated during analysis.	Planned process, organization architecture, and integrative mechanisms provide different views on planned communication and were clearly distinguished.

6.1.5.1 Model of Planned Communication

A process map that shows sequential dependencies between modeling tasks served as the basis for the model of planned communication. Scenario analysis of iteration between tasks helped determine the structure of dependencies between tasks.

The role of BIM developer K partially obscures iteration in the model. BIM developer K executes three modeling tasks and is responsible for resolving clashes between the associated building systems. The three modeling tasks stem from two different batches of the BIM development process. Thus, deduction into the organization domain merges communication regarding building systems from different batches in the model. Merging of communication limits significance of the model. Problem analysis did not focus on BIM developer K's role. Therefore, this limitation of the model did not significantly affect problem analysis.

The method for communication improvement using delta-analysis needs adjustment in order to function properly in projects, which do not have a one-to-one-mapping between people and entities of the indicator domains. Comparison of communication between roles becomes especially necessary in smaller projects, where, for example, one BIM developer takes on several modeling tasks. This adjustment of the method might come with a drawback: during step 9 "Conduct workshop with project team using graphs" people might not identify as well with their role(s) as they would with their own name.

6.1.5.2 Model of Actual Communication

A clash report served as basis for the model of actual communication. Clashes cover only a subset of communication between BIM developers, and several other purposes for communication exist. However, the goal of the planned process was to avoid clashes, so communication regarding clashes is a relevant indicator for communication.

Attributes of the clash report cover only a subset of the characteristics of actual communication. For example, the clash report does not show the severity of each clash and the communication requirements between BIM developers to resolve the clash. In simple cases, clashes may be resolved in a quick conversation between two BIM developers. In more complicated cases, more building systems might be involved and more designers participate in the conversation.

The model of actual communication also omitted an attribute of clashes: it models clashes as binary, i.e., clashes exist between building systems or no clashes exist between building systems. But the number of clashes between systems impacts the necessary communication between BIM developers. Figure 65 shows a force-directed graph of communication between BIM developers based on the clash report that was used in case study B1. This graph models the attribute 'number of clashes between systems' as an integer value, i.e., it takes the number of clashes between systems into account when shaping the graph. The graph was visualized with Gephi (Bastian et al. 2009) using 'Force-Atlas' and 'Label Correct' algorithms. The graph assumes a linear distribution of work between BIM developers working on the same system, i.e., if two BIM developers resolve clashes for one system, each person resolves half of all clashes. Shades of red in entities of the graph represent degree centrality of entities and thickness of lines between entities represent the strength of relations, which indicates communication based on the number of clashes.

Structurally the graph of figure 65 (model includes number of clashes between systems) bears similarities with the graph in figure 59, which does not take into account the number of clashes between systems. Most importantly, BIM developers O and J have a high degree centrality in both networks.

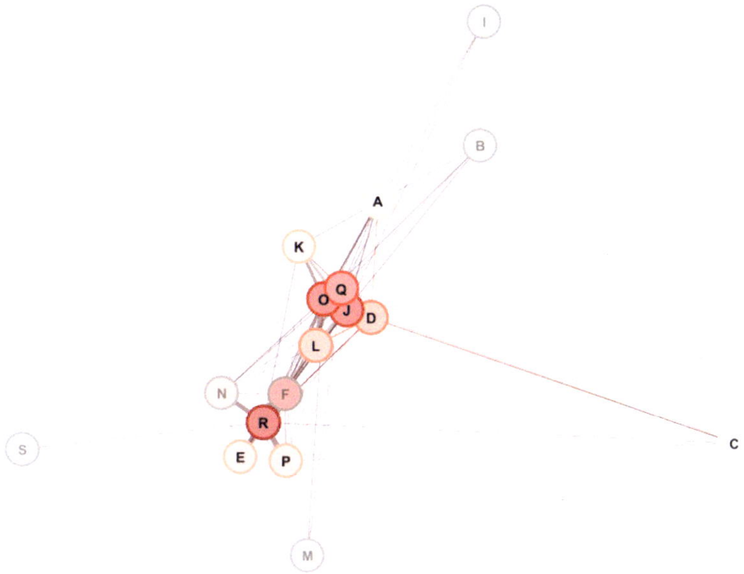

Figure 65: Force-directed Graph of DSM 'communication, as-is' with weighted Relations (Shades of Red in Entities indicate Degree Centrality)

Table 28 lists the degree centralities of BIM developers based on the force directed graph with weighted relations (figure 65).

Table 28: Weighted Degree Centralities of DSM 'communication, as-is'

Entity	**Weighted Degree Centrality**
A	97
B	39
C	10
D	501
E	258.5
F	748
G	0
H	0
I	16
J	936.5
K	242
L	438
M	48
N	266
O	936.5
P	258.5
Q	753
R	1073
S	15

The modeling approach used for the 'as-is' perspective of communication lacked some attributes of actual communication regarding clashes between building systems. Analysis of the weighted network showed that the binary modeling approach and the weighted modeling approach yield similar results. The binary model of clashes between building systems can be considered relevant for the modeling purpose.

6.1.5.3 Visualization of Degree Centrality

Algorithms used in this analysis do not exactly represent degree centrality. Approaches to visualizing centrality in graphs have been proposed (Bannister et al. 2013; Brandes and Wagner 2004) and could be used to improve visualization in future applications.

6.2 Case Study B2 – Large Hospital Project in California BIM Development Process

6.2.1 Case Study Description

The construction project comprises a large hospital in California, USA. Due to confidentiality clauses the author is not allowed to use real names. The project operates under a Guaranteed-Maximum-Price (GMP) contract. The General Contractor, DPR construction, involved builders during the design and preconstruction phase. Staff from seven design firms and 15 construction companies worked part-time from a collocated office. The case study took place during the detailed design phase.

The project team applied Lean Construction Methods to design management. Specifically, the project team used the Last Planner System (Ballard 1994, 2000c) and Target Value Design (Ballard 2011; Ballard and Reiser 2004; Zimina et al. 2012). Also, the project team modeled the hospital facility in BIM.

This case study also applied the software Microsoft Excel for analyzing communication structures.

6.2.2 Modeling of Communication

As in case study B1, case study B2 applies a similar rational for modeling communication: BIM developers want to prevent clashes in the BIM in order to avoid wasteful rework. Similar to case study B1, the modeling process serves as an indicator for planned communication while clashes between building systems serve as indicator for actual communication. Modeling of communication takes place in the organization domain; entities of the organization domain are BIM developers.

6.2.3 Practical Implementation

Practical implementation followed the 10 Step approach outlined in section 5.4.1.2. The software LOOMEO was used to deduce relations and to visualize and analyze graphs.

(1) Kick-off:

The author of this dissertation initiated the improvement project by presenting the method for improvement to the BIM Manager and a BIM Engineer of the project. The project chunks the building into areas that are smaller than floors, and each of the chunks goes through a process of sign-offs. Smaller areas equal small batches of information flowing through the modeling process, which enables shorter project duration.

One sign-off within the process is 'Construction Modeling and Coordination'. The process was not defined at a finer level of granularity, therefore no process model existed which could have served for modeling planned communication.

After presenting the method, BIM Manager, BIM Engineer, and the author developed a BIM modeling process for the current design phase of this project. The process served as the basis for deducing planned communication.

(2) Choose indicators and obtain matrix (-es) for 'should' perspective

In a workshop, BIM Manager, BIM Engineer, and the author decided to use the developed process for modeling planned communication. Similar to case study B1, the modeling process consists of three batches of tasks. The planned process includes iteration within each batch but no iteration between batches. Figure 66 presents a DSM model of the process.

		1	2	3	4	5	6	7	8	9	10	11	12	13	14	15	16	17	18
Model Graded Plumbing (storm drain & sanitary sewer)	1		1																
Model misc. Steel details	2			1															
Model Mechanical Dry Mains (coming from shafts & equipment)	3				1														
Model Mechanical Wet (hydronic piping)	4					1													
Model Drywall – King & Corner Studs	5						1												
Model Drywall – Soffits & Kickers	6							1											
Model Lighting – Public	7								1										
Model Electrical Mains	8									1									
Model Fire Protection Mains & Branch	9										1								
Model Pneumatic Tube	10	1										1							
Model Lighting – General	11																		
Model Med Gas	12													1					
Model Domestic Water	13														1				
Model Cable Tray	14											1				1			
Model Electrical Branch Conduits	15																1		
Model Fire Protection Drops	16																	1	
Model Drywall – Filler Studs	17																		1
Model Electrical Devices	18															1			

Figure 66: DSM 'indicator, should'; Entities represent Modeling Tasks

(3) Obtain indicator-responsibilities DMM for 'should' perspective

The DMM denotes team members' responsibilities for tasks of the BIM development process. Responsibilities were collected from BIM Manger and BIM Engineer of the project during the above described workshop.

Only eight BIM developers conduct 18 modeling tasks, so several BIM developers conduct more than one task. For example, BIM developer F models five systems: lighting-public, electrical mains, lighting-general, cable tray, and electrical devices. Tasks for modeling these five systems stem from all three batches of the BIM development process. A model of communication between BIM developers would merge communication regarding tasks from different batches, and therefore hinder analysis of iteration during delta-analysis of communication. Modeling of roles instead of people in the organization domain mitigates this problem. The DMM 'indicator, should' therefore captures relations between tasks and roles.

(4) Deduce DSM 'communication, should'

Multiplication of DMM 'indicator, should' with DSM 'indicator, should' and transposed DMM 'indicator, should' yielded DSM 'communication, should'. Figure 67 shows DSM 'communication, should'; letters represent roles.

		1	2	3	4	5	6	7	8	9	10	11	12	13	14	15	16	17	18
A	1		1								1								
B	2	1		1															
C	3		1		1														
D	4			1		1													
E	5				1		1												
F	6					1		1											
G	7						1		1										
H	8							1		1									
I	9								1		1								
J	10	1								1		1							
K	11										1				1				
L	12											1		1					
M	13												1		1				
N	14											1		1		1			
O	15														1		1		1
P	16															1		1	
Q	17																1		1
R	18															1		1	

Figure 67: DSM 'communication, should'; Letters represent Roles

(5) Choose indicators and obtain matrix (-es) for 'as-is' perspective

The author, jointly with BIM Manager and BIM Engineer, decided to use clashes between buildings systems as indicators for actual communication. Similar to case study B1, clashes only represent a subset of communication between BIM developers, but attending to clashes is wasteful rework. Figure 68 shows the clash report which shows numbers of clashes between building systems for a specified area of the building.

The clash report summarizes clashes by trade: each trade partner receives an assessment of clashes for their system.

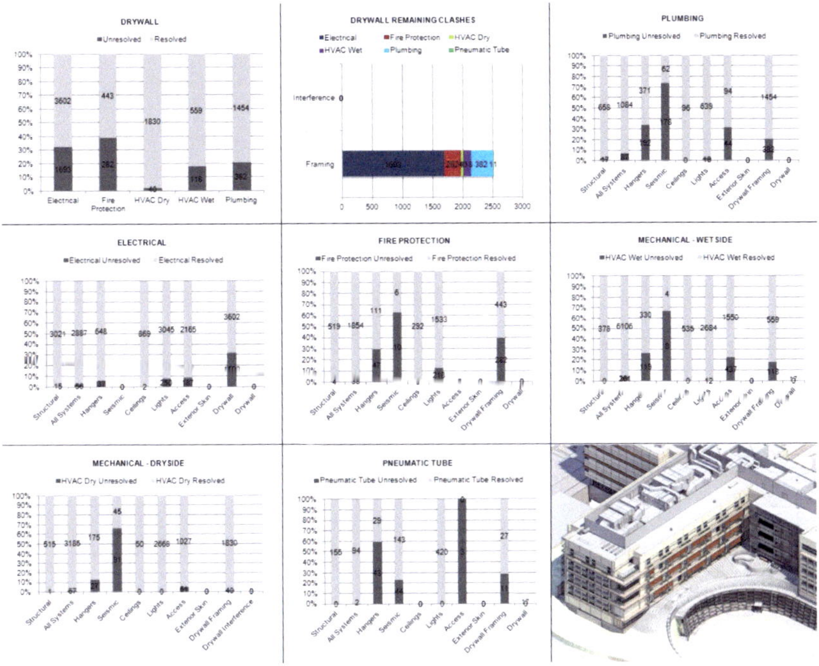

Figure 68: Clash Report from 2011-10-19 (courtesy of DPR Construction)

This setup of the clash report is not applicable to modeling of communication, because it categorizes clashes into only eight categories. These categories do not align with the modeling tasks in DSM 'indicator, should'. Figure 69 shows relations between categories of the clash report and modeling tasks. Tasks from more than one batch relate to one clash category. For example, the system 'electrical' relates to the modeling tasks lighting-public, electrical mains, lighting-general, and electrical devices. A model of communication based on clashes would not allow for an analysis of relations between tasks, because categories of the clash report merge communication that relates to tasks from different batches of the modeling process.

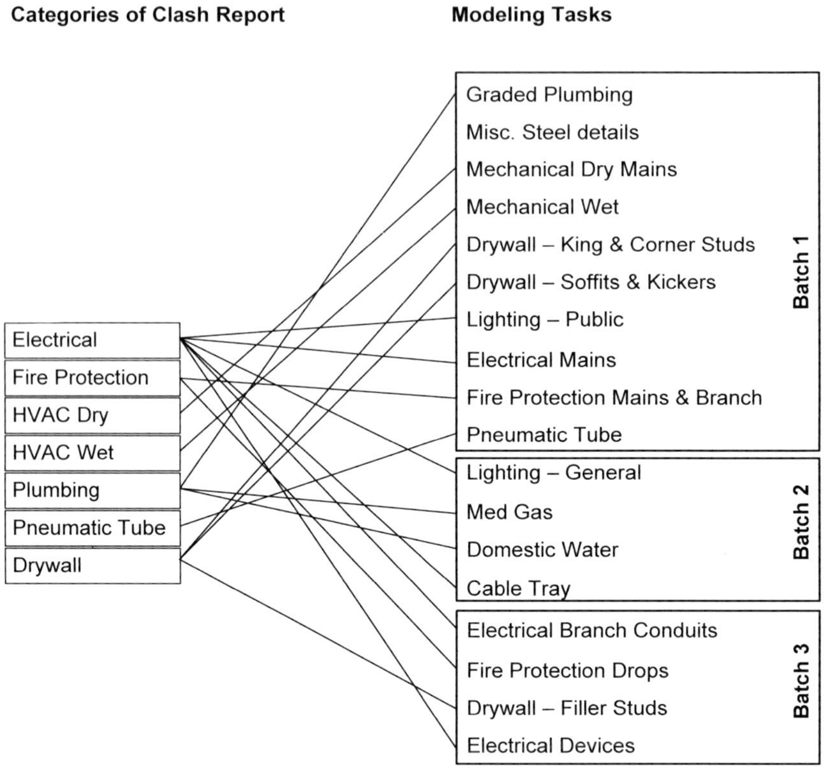

Figure 69: Relations between Categories of Clash Report and Modeling Tasks

Analysis shows that the clash report was not applicable to modeling communication for delta-analysis. Therefore, application of the method for communication improvement using delta-analysis concluded at step 5. The take-aways of applying the method were (1) to restructure the clash report towards a clash matrix, which shows clashes between systems, and (2) to have a larger number of categories which align with modeling tasks. Table 29 summarizes the identified problem and recommended action.

Table 29: Identified Problems and recommended Actions of Case Study B2

Identified Problem	Recommended Action
Clash report groups clashes related to several tasks into one category.	Introduction of a process-oriented clash report.

6.2.4 Results of Application

Analysis of the clash report with BIM Manager and BIM Engineer of the project led to the conclusion that the current structure of the clash report does not support application of the method for communication improvement using delta-analysis. The current clash report lists clashes as responsibilities per project partner, e.g., the category 'electrical' indicates what clashes must be resolved by the BIM developer of the electrical system. Thus, the current structure of the clash report appears related to the structure of contracts between companies instead of being related to the process structure.

Table 30 relates the identified problem with the root domain of the problem and sets the problem in context of lateral relations. The clash report is a tool and as such part of the organization domain of the project. The purpose of the clash report is to achieve situation visibility regarding necessary rework due to clashes in the BIM.

Comparison of the clash report with a modeling process revealed possible improvements to the clash report. So, partial application of the method for communication improvement using delta-analysis yielded relevant results.

Table 30: Identified Problems of Case Study B2 in Context of System Model and Lateral Relations

Identified Problem	Root Domain	Part of lateral relations
Focus and level of detail of clash report do not align with modeling process	Organization - Tools	Integrative mechanisms: situation visibility

6.2.5 Critical Review of Modeling

Case study B2 revealed two prerequisites for successful application of the method for communication improvement using delta-analysis:

(1) The structure and level of detail of 'should' and 'as-is' indicators for communication should be aligned at a similar level of detail.

(2) The structure of relations between (1) people, (2) entities from the 'should'-indictor domain (e.g., tasks), and (3) entities from the 'as-is' indicator domain (e.g., categories of the clash-report) should be similar. For example, one person should at best execute only one modeling task which relates to only one clash category.

Prerequisite 1 can be achieved by restructuring the clash-report. Prerequisite 2 was achieved by changing the modeling approach: BIM developers (being entities of the organization domain) were substituted with their roles in the organization. This type of modeling generates a one-to-one-mapping between modeling tasks and modeling roles.

6.3 Cross-Case Analysis

Cross-case analysis serves to identify similarities and differences between case studies. Table 31 compares important characteristics of the two case studies B1 and B2.

In both case studies the method for communication improvement using delta-analysis was applied during detailed design. In both cases the modeling process served as indicator for planned communication and BIM clashes served as indicator for actual communication. Also, in both case studies problems originated also outside the process domain, for example in the organization and the project environment. The number of BIM developers involved was smaller in case study B2, which took place in a single team environment. The case study B2 project applied a less collaborative contract type.

Table 31: Comparison of Case Studies B1 and B2

	Case Study B1	**Case Study B2**
Indicators for planned communication	BIM development process	BIM development process
Indicators for actual communication	BIM Clash report	BIM Clash report
Number of BIM developers in organization domain	19	8
Workshop approach applied	Yes	No
Project phase	Detailed Design	Detailed Design
Organization architecture of project	Multi-team environment (four cluster groups)	Single team environment
Contract type	IFOA	GMP
Origin of root-causes for identified problems	Process, organization, project environment	Organization

6.4 Limitation of Method

The following limitations of the method for improvement of communication structures using delta-analysis were identified during case studies:

- The method is only applicable when each person is related to only one or very few entities of each indicator domain. For example, in the case of a person executing more than one task of a process (and this process serves as indicator for planned communication), the model of planned communication might become insignificant for analysis of communication structures. In that case, modeling entities in the organization domain based on roles instead of people can increase significance of the model.
- The structure of indicators for actual and planned communication must align; otherwise, comparison of communication structure does not yield significant results.
- The method, as presented in chapter 6, models only binary dependencies, but relations between entities of the domains can be weighted. Models based on binary dependencies can present a 'distorted picture' of indicator domain.
- Some algorithms for visualization of force-directed graphs do not exactly represent degree centralities of entities.

The data which was used in both case studies had limitations. In case study B1 (VNGC), one modeler executed three tasks. In case study B2, the structures of indicators did not align. Nevertheless, both case studies identified opportunities for improvement.

Both case studies used BIM for modeling the 'as-is' perspective on communication, however not all projects use BIM technology. The method elaborated in this dissertation is also applicable to projects which do not use BIM. Other indicators can be obtained for modeling the 'as-is' perspective on communication. For example, emails send between people can serve as an indicator, or a survey can capture actual communication. See section 5.4.2 for an overview of possible data sources for indicators.

6.5 Summary

The method for improvement of communication structures using delta-analysis represents a practical contribution of this dissertation. Application of the method was successful and yielded relevant results in practice. Also, application of the method identified constraints for application and ideas for future improvement of the method.

Use of indicators for modeling communication increased efficiency of the modeling process. Comparison of models for actual and planned communication using force-directed graphs helped in aligning constructed realities of meeting participants. The workshop approach supported identification of root-causes for problems. Foundation of the method in Lean principles of experimentation and investigation aligns with continuous improvement efforts. Thus, it is recommended to apply the method as part of a PDCA cycle to continuously improve the design system.

7 Conclusions

This chapter presents the conclusions of this dissertation. Section 7.1 summarizes the research findings, section 7.2 presents contributions to knowledge, section 7.3 gives recommendations for future work, and section 7.4 closes the dissertation with final remarks.

7.1 Research Findings

7.1.1 Case Study A – Analysis of Communication Structures

Chapter 4 presented case study A, which consists of two parts: (1) a description of the formal organization structure including integrative mechanisms and coordination mechanisms, and (2) a model of the informal organization based on information flow and an analysis of this model using SNA.

Description of the formal organization structure presented the different integrative mechanisms and communication channels. These affect how the informal organization turns out. Hypotheses regarding the informal organization structure were formulated based on relevant literature. These hypotheses were tested with SNA metrics. Results showed that the informal organization possesses structural characteristics which are akin to those prescribed in relevant literature.

Additionally, the communication structure of the VNGC project confirms that this project organization is integrated and flexible. Cluster analysis shows that in all cluster groups designers and builders closely interact, i.e., cross-lifecycle integration exists. Analysis of degree centralities showed that some people take on a coordinating role, even though it is not part of their job description. This finding serves as evidence for flexibility in the organization.

SNA proved useful for identifying information leaders in the design organization. The presented analysis has limitations. In case study A, the author analyzed only one phase of one project, so the significance of results is limited. Also, the model focuses purely on the existence and frequence of communication, and does not include any other attributes regarding content, release of work, batch size of information, or others.

7.1.2 Method for Improvement of Communication Structures Using Delta-Analysis

This section aims at providing answers to the research question and the question regarding impact and procedure (IP) which were formulated in chapter 1.3. The following section presents answers to the research questions:

Q1. How can a design team efficiently achieve transparency of actual and planned communication in the detailed design phase of a construction project?

Answer: Design teams can use indicators to achieve transparency regarding actual communication. In both case studies B1 and B2 BIM clashes served as indicators for actual communication. It must be noted that clash resolution is a wasteful task. However, current industry practice is far from clash-free processes, thus clash resolution is a type of communication worth studying. Processes, integrative mechanisms, and organization architecture can serve as indicators for planned communication

Q2. How can the design team evaluate alignment of actual and planned communication? What are the metrics for evaluation?

Answer: The MDM method combined with delta-DSM serves to detect misalignments. Metrics were presented in chapter 5. Force-directed graphs help in visualizing models of actual and planned communication.

Q3. How can the team use knowledge about misalignments between actual and planned communication to improve the design system continuously?

Answer: Combination of visualization and lean management, especially root-cause analysis, in a workshop setting can help identify opportunities for improvement. Cyclic application of the method as part of a PDCA cycle strengthens continuous improvement efforts.

The following section presents answers to the questions regarding impact and procedure.

IP1. What are the qualitative impacts of application of the method on cost, quality and schedule?

Answer: Application of the method identified root-causes for wasteful rework, which impacts cost and schedule. Elimination of root-causes is expected to affect cost and schedule positively. Quality of the final BIM is not only defined by being clash-free, but also other quality criteria, e.g., well coordinated systems and efficient design. Therefore, additional research regarding the impact on quality is necessary.

IP2. What resources are needed to implement the method?

Answer: Implementation requires data for modeling actual and planned communication, modeling software, manpower for building models, and a workshop for analysis. Data gathering is feasible with existing data sources. These sources can serve as indicators for actual and planned communication. Professional software packages, e.g., LOOMEO, facilitate modeling and analysis. Modeling is also possible using spreadsheet software, e.g., MS Excel, in combination with SNA software, e.g., Gephi (Bastian et al. 2009).

IP3. Who leads method implementation? What skills are necessary for implementation?

Answer: The implementer needs knowledge about processes, goals, and structure of organization and project, e.g., chief engineer or a team leader as they should possess this knowledge. Necessary skills include knowledge of modeling techniques and related software, and skills to guide problem analysis including root-cause analysis.

IP4. What barriers to implementation of the method exist?

Answer: Understanding the method, data gathering and modeling of communication structures are prerequisites for implementation of the method. Also, people's willingness to change behavior is a prerequisite for successful improvement.

7.2 Contributions to Knowledge

The research elaborated in this dissertation contributes to knowledge by providing:

(1) identification of a gap in existing literature regarding proof of existence for prescribed characteristics of IPD projects. Chapter 4

presents literature-based hypotheses for IPD projects, which are then tested,

(2) a social network model of an IPD project design organization. Chapter 4 presents the model whose underlying data was collected by the author through a survey. To the knowledge of the author, it is the first SNA of an informal IPD organization. The contribution to knowledge pertains to model development and analysis of the informal organization,

(3) evidence for the existence of an integrated and flexible project organization in an IPD project design organization at the VNGC project. Analysis of the social network model in chapter 4 yielded evidence for an integrated and flexible project organization. Also, findings support the hypothesis that IPD successfully promotes a 'best-for-project' thinking in the project organization. SNA can serve as a way to visualize and give feedback on the quality of communication structures,

(4) a gap in existing literature regarding post-mortem process evaluation in design by comparison of actual and planned communication. Chapter 3 identified the gap and presented existing methods for process evaluation,

(5) a data gathering method for modeling actual communication in design. Chapter 5 presented the theoretical foundation, which was applied in case studies B1 and B2 of chapter 6,

(6) a method for delta-analysis of actual and planned communication including meta-model and application procedure, which is based in Lean Management. Chapter 5 presented the method which was applied in case studies B1 and B2. Case study B1 and B2 showed that comparison of models of actual and planned communication has practical relevance in the AEC industry. Results show that post mortems of prescriptive processes can identify opportunities for improvement,

(7) a new use-case for BIM as a data source for process modeling. Case studies B1 and B2 showed that logs of the BIM database can be used for modeling actual communication between BIM users. Case study B1 (VNGC) showed that BIM clashes can serve as an indicator for actual communication between BIM developers. This finding is important, because data gathering through databases takes less effort than data gathering through surveys.

These seven areas of contribution provide a foundation for further discussion of communication structures in AEC design projects. Extensions to the research that has been elaborated in this dissertation will be discussed next.

7.3 Recommendations for Future Work

Completion of the case studies raised a number of questions that remain for future work.

- Organization design of IPD projects seems a promising field for future research. Moving away from the traditional silo-structure, IPD projects enable new ways for architecting the organization. This dissertation identified several questions for future research: How can social network models of design organizations be amended to include additional information regarding content and release of work to others? The models presented in this research focus purely on the existence and frequence of communication. Flores' LAP could be applied in conjunction with SNA to model hand-offs and content.
- How can SNA be used as a diagnostic tool? How can be identified whether a person is fulfilling their role? How can he/she be empowered to fulfill their role? Application of SNA on more projects and in different phases will help to build a frame of reference for comparison of organizational structures between different project phases and between projects. Also, comparison of communication

networks of IPD and non-IPD projects will most probably yield interesting insights.

- Why is information flow between designers un-evenly distributed in case study A? Are there advantages to different types of distributions of information flow? Again, application of SNA on more projects and in different phases will help to build a frame of reference.
- How can flexibility of project organizations be evaluated in longitudinal studies of information flow over time? These studies would have higher significance in terms of the assessment of flexibility.
- How can clash-free modeling processes be achieved? Clash resolution is a wasteful task and should be avoided. The modeling sequence seems to play an important role in avoiding clashes.
- How can models of communication structures based on weighted relations be deduced and compared with delta-analysis? The method presented in this research can be extended by modeling and comparing weighted communication-type relations.
- How can process data gathering in design be extended to capture additional attributes of interaction between people? How can data be collected that serves for modeling specific communication channels? How can data gathering be extended to capture additional attributes of ties, e.g., mode of information transfer (push/pull), batch size of information transfer, and processing times? This data could be used to build current state VSMs of the design process.
- How should the method elaborated in this research be used? When and how often should it be applied? What are the impacts of the method on cost and time? How can validity of models built from indicators be checked? Additional studies and application of the method in recurring PDCA cycles will expand knowledge about utilization of the method.
- How can the method presented in this research be transferred to other phases of the design process? Detailed design, at least in the

case studies of this research, provided BIM and all needed people on-board the project. How can the method be applied under different circumstances?

7.4 Final Remarks

This research focuses on communication and the related structures in project organizations. The research expands previous knowledge about communication structures in IPD projects, and this research expands previous knowledge about methods for improving communication structures.

Three different case studies showed the importance of transparent communication structures for design process management. Case studies also showed the importance of reflection of communication structures for team learning. This research focused on (1) reducing effort for making communication structures transparent and (2) on applying the concepts of investigation and experimentation to the design process. Regarding (1), use of indicators for communication has been identified as applicable to engineering design. Regarding (2), involvement of process, stakeholders, visualization, and use of the scientific method have proven successful in this research. Specifically, the scientific method and root-cause analysis have proven to be powerful methods for installing an open systems perspective on the design process during improvement efforts. The open systems perspective is a key lever for making design process post-mortems effective, because learnings about reasons for deviating from planned processes can originate outside the process domain. Even though the specific design process is not being repeated, reasons for deviations can persist and learning about them can help in improving subsequent design processes.

Finally, the method presented in this dissertation can be further studied in order to expand its range of utilization to other project types and design phases and to learn about how and when to use it. Expanding the concept of using indicators for modeling can reduce the effort for achieving transparency regarding actual communication. However, such efforts may conflict with concerns of individuals regarding their privacy. Such concerns shall be taken seriously.

The presented modeling approach and use of SNA in project management are first steps. Increased use of information technology and digitization will facilitate data gathering and thereby boost opportunities for modeling, analysis, and improvement of production systems.

References

Aalst, W. M. van der. (2005). "Business alignment: using process mining as a tool for Delta analysis and conformance testing." *Requirements Engineering*, 10(3), 198–211.

Aalst, W. M. van der. (2011). *Discovery, Conformance and Enhancement of Business Processes*. Springer, Heidelberg, Germany.

Albers, A., and Meboldt, M. (2007). "SPALTEN Matrix—Product Development Process on the Basis of Systems Engineering and Systematic Problem Solving." *The Future of Product Development: Proceedings of the 17th CIRP Design Conference*, F.-L. Krause, ed., Springer, Berlin, Germany, 43–52.

Albert, R., Jeong, H., and Barabási, A.-L. (2000). "Error and attack tolerance of complex networks." *Nature*, 406(6794), 378–382.

Alexander, C. (1964). *Notes on the Synthesis of Form*. Harvard University Press, Cambridge, USA.

Allen, T. J. (1977). *Managing the Flow of Technology: Technology Transfer and the Dissemination of Technological Information with the R&D Organization*. MIT Press, Cambridge, MA.

American Institute of Architects. (2007). "Integrated Project Delivery: A Guide." <http://info.aia.org/siteobjects/files/ipd_guide_2007.pdf> (May 30, 2015).

Anumba, C. J., and Evbuomwan, N. F. O. (1997). "Concurrent engineering in design-build projects." *Construction Management and Economics*, 15(3), 271–281.

Argyris, C., and Schön, D. A. (1978). *Organizational learning: a theory of action perspective*. Addison-Wesley Pub. Co., Boston, USA.

Ashby, W. (1973). "Some peculiarities of complex systems." *Cybernetic Medicine*, 9, 1–7.

Ashby, W. R. (1956). *An Introduction to Cybernetics*. Taylor & Francis, New York, USA.

Austin, S., Baldwin, A., Li, B., and Waskett, P. (2000). "Analytical design planning technique (ADePT): a dependency structure matrix tool to schedule the building design process." *Construction Management & Economics*, 18(2), 173–182.

Baccarini, D. (1996). "The concept of project complexity–a review." *International Journal of Project Management*, 14(4), 201–204.

Baecker, D. (2003). *Organisation und Management: Aufsätze*. Suhrkamp, Frankfurt, Germany.

Baecker, D. (2006). "The Form of the Firm." *Organization*, 13(1), 109–142.

Bahrami, H., and Evans, S. (2011). "Super-flexibility for real-time adaptation: perspectives from Silicon Valley." *California Management Review*, 53(3), 21–39.

Baldwin, C. Y., and Clark, K. B. (2000). *Design rules: The power of modularity*. The MIT Press, Cambridge, USA.

Ballard, G. (1994). "The last planner." *Northern California Construction Institute, Monterey, USA*.

Ballard, G. (1999). "Work Structuring." *White paper of the Lean Construction Institute*, number 5.

Ballard, G. (2000a). "Lean project delivery system." *White paper of the Lean Construction Institute*, number 8.

Ballard, G. (2000b). "Positive vs negative iteration in design." *Proceedings 8th Annual Conference of the International Group for Lean Construction (IGLC)*, Brighton, UK.

Ballard, G. (2000c). "The last planner system of production control." Dissertation, The University of Birmingham, Birmingham, UK.

Ballard, G. (2002). "Managing work flow on design projects: a case study." *Engineering Construction and Architectural Management*, 9(3), 284–291.

Ballard, G. (2008). "The lean project delivery system: an update." *Lean Construction Journal*, 2008, 1–19.

Ballard, G. (2011). "Target value design: Current benchmark (1.0)." *Lean Construction Journal*, 2011, 79–84.

Ballard, G. (2012). "Should Project budgets be based on worth or cost." *Proceedings 20th Annual Conference of the International Group for Lean Construction (IGLC)*, San Diego, USA, 761–770.

Ballard, G., and Koskela, L. (1998). "On the agenda of design management research." *Proceedings 6th Annual Conference of the International Group for Lean Construction (IGLC)*, Guarujá, Brazil.

Ballard, G., Koskela, L., Howell, G., and Zabelle, T. (2001a). "Production system design: Work structuring revisited." *White paper of the Lean Construction Institute*, number 11.

Ballard, G., Koskela, L., Howell, G., and Zabelle, T. (2001b). "Production system design in construction." *Proceedings 9th Annual Conference of the International Group for Lean Construction (IGLC)*, Singapore, Singapore.

Ballard, G., and Reiser, P. (2004). "The St. Olaf College Fieldhouse Project: a case study in designing to target cost." *Proceedings 12th Annual Conference of the International Group for Lean Construction (IGLC)*, Helsingør, Denmark, 234–49.

Ballard, G., and Tommelein, I. (2012). "Lean management methods for complex projects." *Engineering Project Organization Journal*, 2(1-2), 85–96.

Bannister, M. J., Eppstein, D., Goodrich, M. T., and Trott, L. (2013). "Force-Directed graph drawing using social gravity and scaling." *Graph Drawing*, 414–425.

Barabási, A.-L. (2015). *Network Science*. Cambridge University Press.

Barabási, A.-L., and Bonabeau, E. (2003). "Scale-free networks." *Scientific American*, 288(5), 50–9.

Bashir, H. A., and Thomson, V. (1999). "Metrics for design projects: a review." *Design Studies*, 20(3), 263–277.

Bastian, M., Heymann, S., and Jacomy, M. (2009). "Gephi: an open source software for exploring and manipulating networks." *ICWSM*, Seattle, USA, 361–362.

Battista, G. D., Eades, P., Tamassia, R., and Tollis, I. G. (1998). *Graph Drawing: Algorithms for the Visualization of Graphs*. Prentice Hall PTR, Upper Saddle River, NJ, USA.

Bauch, C. (2004). "Lean Product Development: Making waste transparent." Diploma Thesis, Massachusetts Institute of Technology, Cambridge, USA.

Becker, J., Rosemann, M., and Schütte, R. (1995). "Grundsätze ordnungsmäßiger Modellierung." *Wirtschaftsinformatik*, 37(5), 435-445.

Benbasat, I., Goldstein, D. K., and Mead, M. (1987). "The Case Research Strategy in Studies of Information Systems." *MIS Quarterly*, 11(3), 369.

Bergsjö, D., Vielhaber, M., Malvius, D., Burr, H., and Malmqvist, J. (2007). "Product lifecycle management for cross-x engineering design." *Proceedings International Conference on Engineering Design (ICED07), Paris, France.*

Bertalanffy, L. V. (1950). "An outline of general system theory." *British Journal for the Philosophy of Science*, 1, 134–165.

Bertelsen, S. (2003). "Complexity–Construction in a new Perspective." *Proceedings 11th Annual Conference of the International Group for Lean Construction (IGLC)*, Blacksburg, USA.

Birrell, G. S. (1981). "The Informal Organization which Manages the Construction Process." *Proc. CIB W-65 3rd Symp. on Organ. and Mgmt. of Construction*, Dublin, Ireland, 201–211.

Blessing, L., and Chakrabarti, A. (2009). *DRM: A Design Research Methodology*. Springer, London , UK.

Both, P. von. (2006). "Ein systemisches Projektmodell für eine kooperative Planung komplexer Unikate." Unversität Karlsruhe, Karlsruhe, Germany.

Both, P. von. (2011). "Produktdatenmodellierung–aktuelle Entwicklungen und Möglichkeiten der Vernetzung von Produkt-und Prozessebene." *Zukunftspotenzial Bauwirtschaft, Proc. 1. Int. BBB-Kongress*, 131.

Braha, D., and Bar-Yam, Y. (2004). "Topology of large-scale engineering problem-solving networks." *Physical Review E*, 69(1), 016113.

Brandes, U., and Wagner, D. (2004). "Analysis and visualization of social networks." *Graph drawing software*, Springer, 321–340.

Browning, T. R. (1998). "Use of dependency structure matrices for product development cycle time reduction." *Proceedings of the Fifth ISPE International Conference on Concurrent Engineering: Research and Applications*, Tokyo, Japan, 89–96.

Browning, T. R. (2001). "Applying the design structure matrix to system decomposition and integration problems: a review and new directions." *IEEE Transactions on Engineering Management*, 48(3), 292–306.

Browning, T. R. (2002). "Process integration using the design structure matrix." *Systems Engineering*, 5(3), 180–193.

Browning, T. R. (2003). "On customer value and improvement in product development processes." *Systems engineering*, 6(1), 49–61.

Browning, T. R. (2009). "Using the Design Structure Matrix to design program organizations." *Handbook of Systems Engineering and Management*, A. P. Sage and W. B. Rouse, eds., Wiley, New York, USA.

Browning, T. R., and Eppinger, S. D. (2002). "Modeling impacts of process architecture on cost and schedule risk in product development." *IEEE Transactions on Engineering Management*, 49(4), 428–442.

Browning, T. R., Fricke, E., and Negele, H. (2006). "Key concepts in modeling product development processes." *Systems Engineering*, 9(2), 104–128.

Browning, T. R., and Ramasesh, R. V. (2007). "A Survey of Activity Network-Based Process Models for Managing Product Development Projects." *Production and Operations Management*, 16(2), 217–240.

Burns, T. E., and Stalker, G. M. (1961). *The management of innovation*. Tavistock, London, UK.

Burt, R. S. (2004). "Structural Holes and Good Ideas." *American Journal of Sociology*, 110(2), 349–399.

Carrillo, P. (2005). "Lessons learned practices in the engineering, procurement and construction sector." *Engineering, Construction and Architectural Management*, 12(3), 236–250.

Cataldo, M., Wagstrom, P. A., Herbsleb, J. D., and Carley, K. M. (2006). "Identification of coordination requirements: implications for the Design of collaboration and awareness tools." *Proceedings of the 20th anniversary conference on Computer supported cooperative work*, ACM, Banff, Canada, 353–362.

Chandler, D., Hickethier, G., Lostuvali, B., and Sparapani, A. (2011). "PDCA Thinking In Design and Trade Integration." Presentation 13th Annual LCI Congress, Pasadena, USA.

Checkland, P. B. (1989). "Soft Systems Methodology." *Human Systems Management*, 8(4), 273–289.

Chinowsky, P., Diekmann, J., and Galotti, V. (2008). "Social Network Model of Construction." *Journal of Construction Engineering and Management*, 134(10), 804–812.

Chinowsky, P., Taylor, J., and Di Marco, M. (2011). "Project Network Interdependency Alignment: New Approach to Assessing Project Effectiveness." *Journal of Management in Engineering*, 27(3), 170–178.

Cho, S., and Ballard, G. (2011). "Last planner and integrated project delivery." *Lean Construction Journal*, 7(1), 67–78.

Chu, D., Strand, R., and Fjelland, R. (2003). "Theories of complexity." *Complexity*, 8(3), 19–30.

CII. (2011). *Transforming Modular Construction for the Competitive Advantage through the Adaptation of Shipbuilding Production Processes to Construction*. Research Summary, Construction Industry Institute (CII), Austin, USA.

Clarkson, J., and Eckert, C. (2005). *Design Process Improvement: A review of current practice*. Springer, London, UK.

Colfer, L., and Baldwin, C. Y. (2010). *The mirroring hypothesis: Theory, evidence and exceptions*. Harvard Business School Working Paper, Harvard Business School, Cambridge, USA.

Conklin, E. J., and Weil, W. (1997). "Wicked problems: naming the pain in organizations." Reading Room Research Paper.

Court, P. F. (2009). "Transforming traditional mechanical and electrical construction to a modern process of assembly." Thesis, Loughborough University, Loughborough, UK.

Cristiano, J. J., Liker, J. K., and White III, C. C. (2001). "Key factors in the successful application of quality function deployment (QFD)." *IEEE Transactions on Engineering Management*, 48(1), 81–95.

Cross, N. (1984). *Developments in design methodology*. Wiley, Hoboken, USA.

Cushman, R. F., and Loulakis, M. C. (2001). *Design-build Contracting Handbook*. Aspen Publishers, New York, USA.

Da Rocha, C. G., Formoso, C. T., Tzortzopoulos-Fazenda, P., Koskela, L., and Tezel, A. (2012). "Design Science Research in Lean Construction: Process and Outcomes." *Proceedings 20th Annual Conference of the International Group for Lean Construction (IGLC)*, San Diego, USA.

Danilovic, M., and Browning, T. (2004). "A formal approach for domain mapping matrices (DMM) to complement design structure matrices (DSM)." *Proceedings of the 6th international design structure matrix (DSM) workshop*, Cambridge, UK.

Delgado-Hernandez, D. J., Bampton, K. E., and Aspinwall, E. (2007). "Quality function deployment in construction." *Construction Management and Economics*, 25(6), 597–609.

Deming, W. E. (2000). *Out of the Crisis*. The MIT Press, Cambridge, USA.

Dodgson, M. (1993). "Organizational learning: a review of some literatures." *Organization studies*, 14(3), 375–394.

Donaldson, L. (1999). "The normal science of structural contingency theory." *Studying Organizations: Theory and Method*, S. R. Clegg and C. Hardy, eds., Sage, Thousand Oaks, USA, 51–70.

Dooley, K. J. (1997). "A complex adaptive systems model of organization change." *Nonlinear Dynamics, Psychology, and Life Sciences*, 1(1), 69–97.

Dubois, A., and Gadde, L.-E. (2002). "The construction industry as a loosely coupled system: implications for productivity and innovation." *Construction Management and Economics*, 20(7), 621–631.

Eastman, C. M., Teicholz, P., Sacks, R., and Liston, K. (2008). *BIM handbook: A guide to building information modeling for owners, managers, architects, engineers, contractors, and fabricators*. John Wiley and Sons, Hoboken, USA.

Eben, K., Biedermann, W., and Lindemann, U. (2008). "Modeling Structural Change over Time–Requirements and First Methods." *Proceedings 10th international Design Structure Matrix Conference*, Stockholm, Sweden.

Eckert, C., Clarkson, P., and Stacey, M. (2001). "Information flow in engineering companies: problems and their causes." *Proceedings International Conference on Engineering Design (ICED01)*, Glasgow, UK, 43–50.

Eckert, C. M., and Stacey, M. K. (2010). "What is a Process Model? Reflections on the Epistemology of Design Process Models." *Modelling and Management of Engineering Processes*, P. Heisig, P. J. Clarkson, and S. Vajna, eds., Springer London, 3–14.

Edmonds, B. (1999). "Syntactic measures of complexity." University of Manchester, Manchester, UK.

Eisenhardt, K. M. (1989). "Building theories from case study research." *Academy of management review*, 14(4), 532–550.

Elezi, F., Graebsch, M., and Lindemann, U. (2010). "Reducing Waste in Product Development by Use of Multi-Domain Matrix Methodology." *Proceedings of the 11th International Design Conference DESIGN 2010*, Dubrovnik, Croatia, 205–212.

Engwall, M. (2003). "No project is an island: linking projects to history and context." *Research Policy*, 32(5), 789–808.

Eppinger, S. D. (2001). "Innovation at the Speed of Information." *Harvard Business Review*, 79(1), 149–158.

Eppinger, S. D., and Browning, T. R. (2012). *Design structure matrix methods and applications*. MIT press, Cambridge, USA.

Eppinger, S., and Salminen, V. (2001). "Patterns of product development interactions." *Proceedings International Conference on Engineering Design (ICED01)*, Glasgow, UK.

Eppler, M. J., and Mengis, J. (2003). *A framework for information overload research in organizations*. Insights from Organization Science, Accounting, Marketing, MIS, and Related Disciplines, Università della Svizzera italiana, 2003.

Flores, C. F. (1981). "Management and Communication in the Office of the Future." University of California Berkeley, Berkeley, USA.

Freire, J., and Alarcón, L. F. (2002). "Achieving lean design process: improvement methodology." *Journal of Construction Engineering and Management*, 128(3), 248 - 256.

Fruchterman, T. M. J., and Reingold, E. M. (1991). "Graph drawing by force-directed placement." *Software: Practice and Experience*, 21(11), 1129–1164.

Galbraith, J. R. (1971). "Matrix organization designs How to combine functional and project forms." *Business Horizons*, 14(1), 29–40.

Galbraith, J. R. (1974). "Organization design: An information processing view." *Interfaces*, 4(3), 28–36.

Gehbauer, F. (2008). "„Lean organization: Exploring extended potentials of the last planner system."." *Proceedings of the 16th Annual Conference of the International Group for Lean Construction, Manchester, UK*, 3–13.

Geraldi, J., Maylor, H., and Williams, T. (2011). "Now, let's make it really complex (complicated): A systematic review of the complexities of projects." *International Journal of Operations & Production Management*, 31(9), 966–990.

Gidado, K. I. (1996). "Project complexity: The focal point of construction production planning." *Construction Management and Economics*, 14(3), 213–225.

Gil, N., Tommelein, I. D., Kirkendall, R. L., and Ballard, G. (2000). "Contribution of specialty contractor knowledge to early design." *Proceedings 8th Annual Conference of the International Group for Lean Construction (IGLC)*, Brigthon, UK.

Goldratt, E. M. (1990). *Theory of Constraints*. North-River Press, Croton-on-Hudson, USA.

Granovetter, M. S. (1973). "The strength of weak ties." *American journal of sociology*, 1360–1380.

Gross, J. L., and Yellen, J. (2005). *Graph Theory and Its Applications*. CRC Press, Baco Raton, USA.

Gulati, R. K., and Eppinger, S. D. (1996). *The coupling of product architecture and organizational structure decisions*. Working Paper, Massachusetts Institute of Technology, Dept. of Electrical Engineering and Computer Science, Cambridge, USA.

Gunasekaran, A., and Love, P. E. D. (1998). "Concurrent engineering: a multi-disciplinary approach for construction." *Logistics Information Management*, 11(5), 295–300.

Hammer, M., and Champy, J. (1999). *Reengineering the Corporation: A Manifesto for Business Revolution*. HarperBus, New York, USA.

Hammond, J., Choo, H. J., Austin, S., Tommelein, I. D., and Ballard, G. (2000). "Integrating design planning, scheduling, and control with Deplan." *Proceedings 8th Annual Conference of the International Group for Lean Construction (IGLC)*, Brighton, UK.

Hamzeh, F. R., Ballard, G., and Tommelein, I. D. (2009). "Is the Last Planner System applicable to design?—A case study." *Proceedings 17th Annual Conference of the International Group for Lean Construction (IGLC)*, Taipei, Taiwan.

Hansen, M. T. (1999). "The search-transfer problem: The role of weak ties in sharing knowledge across organization subunits." *Administrative science quarterly*, 44(1), 82–111.

Hatchuel, A., and Weil, B. (2003). "A new approach of innovative design: an introduction to CK theory." *Proceedings of the 14th International Conference on Engineering Design (ICED 03)*, Stockholm, Sweden.

Heidemann, A., and Gehbauer, F. (2010). "Cooperative project delivery in an environment of strict design-bid-build tender regulations." *Proceedings 18th Annual Conference of the International Group for Lean Construction (IGLC)*, Haifa, Israel.

Hellenbrand, D. (2013). "Transdisziplinäre Planung und Synchronisation mechatronischer Produktentwicklungsprozesse." Universität München, München, Germany.

Henderson, R. M., and Clark, K. B. (1990). "Architectural innovation: The reconfiguration of existing product technologies and the failure of established firms." *Administrative science quarterly*, 9–30.

Hevner, A. R., March, S. T., Park, J., and Ram, S. (2004). "Design Science in Information Systems Research." *MIS Q.*, 28(1), 75–105.

Hickethier, G., Anbergen, H., Hofacker, A., and Gehbauer, F. (2011a). "Set-based Planning in the Decommissioning of a Nuclear Power Plant." *Proceedings 19th Annual Conference of the International Group for Lean Construction (IGLC)*, Lima, Peru.

Hickethier, G., Tommelein, I. D., and Gehbauer, F. (2012). "Reducing rework in design by comparing structural complexity using a Multi Domain Matrix." *Proceedings 20th Annual Conference of the International Group for Lean Construction (IGLC)*, San Diego, USA.

Hickethier, G., Tommelein, I. D., Hofmann, M., Lostuvali, B., and Gehbauer, F. (2011b). "MDM as a Tool to Improve BIM Development Processes." *Proceedings of the 13th International DSM Conference*, Cambridge, MA.

Hickethier, G., Tommelein, I. D., and Lostuvali, B. (2013). "Social Network Analysis of Information Flow in an IPD-Project Design Organization." *Proceedings 21th Annual Conference of the International Group for Lean Construction (IGLC)*, Curitiba, Brazil.

Horgan, J. (1995). "From Complexity to Perplexity." *Scientific American*, 272, 104–109.

Howell, G. A. (1999). "What is lean construction-1999." *Proceedings 7th Annual Conference of the International Group for Lean Construction (IGLC)*, Berkeley, USA.

Howell, G., and Ballard, G. (1997). "Lean production theory: Moving beyond 'Can-Do.'" *Lean construction*, L. Alarcon, ed., Balkema, Rotterdam, Netherlands, 17–24.

Howell, G., Ballard, G., and Tommelein, I. (2011). "Construction Engineering—Reinvigorating the Discipline." *Journal of Construction Engineering and Management*, 137(10), 740–744.

Howell, G., Laufer, A., and Ballard, G. (1993). "Uncertainty and project objectives." *Project Appraisal*, 8(1), 37–43.

Howell, G., and Macomber, H. (2002). "A guide for new users of the Last Planner System - nine steps for success." Lean Project Consulting, Inc.

Howell, G., Macomber, H., Koskela, L., and Draper, J. (2004). "Leadership and project management: time for a shift from Fayol to Flores." *Proceedings 12th Annual Conference of the International Group for Lean Construction (IGLC)*, Helsingør, Denmark, 22–29.

Huovila, P., Koskela, L., and Lautanala, M. (1997). "Fast or concurrent: the art of getting construction improved." *Lean construction*, L. Alarcon, ed., Balkema, Rotterdam, Netherlands, 143–159.

Jensen, P., Hamon, E., and Olofsson, T. (2009). "Product development through lean design and modularization principles." *Proceedings 17th Annual Conference of the International Group for Lean Construction (IGLC)*, Taipei, Taiwan.

Jin, Y., and Levitt, R. E. (1996). "The virtual design team: A computational model of project organizations." *Computational & Mathematical Organization Theory*, 2(3), 171–195.

Jungwirth, D., and Fuhr, H. (1994). *Qualitätsmanagement im Bauwesen*. VDI-Verlag, Düsseldorf, Germany.

Kahn, K. B., and McDonough, E. F. (1997). "An Empirical Study of the Relationships among Co-location, Integration, Performance, and Satisfaction." *Journal of Product Innovation Management*, 14(3), 161–178.

Kirsch, J. (2009). "Organisation der Bauproduktion nach dem Vorbild industrieller Produktionssysteme: Entwicklung eines Gestaltungsmodells eines ganzheitlichen Produktionssystems für den Bauunternehmer." Universität Karlsruhe, Karlsruhe, Germany.

Klir, G. J. (1985). "Complexity: some general observations." *Systems Research*, 2(2), 131–140.

Kobourov, S. G. (2013). "Force-Directed Drawing Algorithms." *Handbook of Graph Drawing and Visualization*, CRC Press, 383–408.

Koners, U., and Goffin, K. (2007). "Learning from Postproject Reviews: A Cross-Case Analysis." *Journal of Product Innovation Management*, 24(3), 242–258.

Koskela, L. (1992). *Application of the new production philosophy to construction*. Center for Integrated Facility Engineering, Department of Civil Engineering, Stanford University, Stanford, USA.

Koskela, L. (2000). "An exploration towards a production theory and its application to construction." VTT Technical Research Centre of Finland, Espoo, Finland.

Koskela, L., and Alarcon, L. (1997). "Lean production in construction." *Lean Construction*, L. Alarcon, ed., Balkema, Rotterdam, Netherlands, 1–9.

Koskela, L., Ballard, G., and Tanhuanpaa, V.-P. (1997). "Towards lean design management." *Proceedings 5th Annual Conference of the International Group for Lean Construction (IGLC)*, Gold Coast, Australia.

Koskela, L. J. (2004). "Making do-the eighth category of waste." *Proceedings 12th Annual Conference of the International Group for Lean Construction (IGLC)*, Helsingør, Denmark.

Koskela, L. J., Ballard, G., Howell, G., and Tommelein, I. (2002). "The foundations of lean construction." *Design and construction: building in value*, R. Best and G. de Valence, eds., Butterworth Heinemann, Oxford, UK, 211–226.

Koskela, L. J., and Howell, G. (2002). "The underlying theory of project management is obsolete." *Proceedings of the PMI Research Conference*, Seattle, USA, 293–301.

Koskela, L. J., and Kagioglou, M. (2006a). "The proto-theory of design: the method of analysis of the ancient geometers." *Proceedings DESIGN 2006, the 9th International Design Conference*, Dubrovnik, Croatia.

Koskela, L. J., and Kagioglou, M. (2006b). "On the metaphysics of management." *Proceedings 14th Annual Conference of the International Group for Lean Construction (IGLC)*, Santiago, Chile.

Kratzer, J., Gemünden, H. G., and Lettl, C. (2008). "Balancing creativity and time efficiency in multi-team R&D projects: the alignment of formal and informal networks." *R&D Management*, 38(5), 538–549.

Kreimeyer, M., Eichinger, M., and Lindemann, U. (2007). "Aligning multiple domains of design processes." *16th International Conference on Engineering Design*, Paris, France.

Kreimeyer, M. F. (2009). "A structural measurement system for engineering design processes." Technische Universität München, München, Germany.

Krinner, M., Elezi, F., Tommelein, I. D., and Lindemann, U. (2011). "Managing Complexity in Lean Construction Design–Using the MDM Methodology to Create Organizational Modularity." *Proceedings of the 13th International DSM Conference*, Cambridge, USA.

Kurtz, C. F., and Snowden, D. J. (2003). "The new dynamics of strategy: Sense-making in a complex and complicated world." *IBM systems journal*, 42(3), 462–483.

Lahdenperä, P. (2012). "Making sense of the multi-party contractual arrangements of project partnering, project alliancing and integrated project delivery." *Construction Management & Economics*, 30(1), 57–79.

Lano, R. (1977). *The N2 chart*. Informational Report, TRW, California, USA.

Lawrence, P. R., Lorsch, J., and Garrison, J. S. (1967). *Organization and Environment*. Division of Research, Graduate School of Business Administration, Harvard University, Cambridge, USA.

Lawrence, P. R., and Lorsch, J. W. (1967). "Differentiation and integration in complex organizations." *Administrative science quarterly*, 1–47.

Lennartsson, M., Björnfot, A., and Stehn, L. (2008). "Lean Modular Design: Value Based Progress in Industrialised Housing." *Proceedings 16th Annual Conference of the International Group for Lean Construction (IGLC)*, Manchester, UK, 541–552.

Levin, D. Z., and Cross, R. (2004). "The strength of weak ties you can trust: The mediating role of trust in effective knowledge transfer." *Management science*, 50(11), 1477–1490.

Lichtig, W. A. (2005). "Sutter health: Developing a contracting model to support lean project delivery." *Lean Construction Journal*, 2(1), 105–112.

Lichtig, W. A. (2006). "Integrated Agreement for Lean Project Delivery, The." *Constr. Law.*, 26, 25.

Liker, J. K. (2004). *The Toyota Way*. McGraw-Hill, New York, USA.

Lindemann, U. (2009). *Methodische Entwicklung technischer Produkte: Methoden flexibel und situationsgerecht anwenden*. Springer Verlag, München, Germany.

Lindemann, U., Maurer, M., and Braun, T. (2008). *Structural complexity management: an approach for the field of product design*. Springer Verlag, München, Germany.

Lostuvali, B., Alves, T., and Modrich, R.-U. (2012). "Lean Product Development at Cathedral Hill Hospital Project." *Proceedings 20th Annual Conference of the International Group for Lean Construction (IGLC)*, San Diego, USA.

Love, P. E. D., and Gunasekaran, A. (1997). "Concurrent Engineering in the Construction Industry." *Concurrent Engineering*, 5(2), 155–162.

Love, P. E. ., Edwards, D. J., and Irani, Z. (2008). "Forensic project management: An exploratory examination of the causal behavior of design-induced rework." *IEEE Transactions on Engineering Management*, 55(2), 234–247.

Lukka, K. (2003). "The constructive research approach." *Case study research in logistics. Publications of the Turku School of Economics and Business Administration, Series B*, 1(2003), 83–101.

Macomber, H., and Howell, G. (2003). "Linguistic Action: Contributing to the theory of lean construction." *Proceedings 11th Annual Conference of the International Group for Lean Construction (IGLC)*, Virginia, USA.

Macomber, H., and Howell, G. A. (2004). "Two Great Wastes in Organizations–A Typology for Adressing the Concern for the Underutilization of Human Potential." *Proceedings 12th Annual Conference of the International Group for Lean Construction (IGLC)*, 2007.

Macomber, H., Howell, G. A., and Reed, D. (2005). "Managing Promises with the Last Planner System: Closing in on Uninterrupted Flow." *Proceedings 13th Annual Conference of the International Group for Lean Construction (IGLC)*, Sydney, Australia.

Maier, A. M., Kreimeyer, M., Hepperle, C., Eckert, C. M., Lindemann, U., and Clarkson, P. J. (2008). "Exploration of Correlations between Factors Influencing Communication in Complex Product Development." *Concurrent Engineering*, 16(1), 37–59.

Maier, A., and Störrle, H. (2011). "What are the characteristics of engineering design processes." *Proceedings 18th International Conference on Engineering Design (ICED)*, Copenhagen, Denmark, 188–198.

Malone, T. W., and Crowston, K. (1990). "What is coordination theory and how can it help design cooperative work systems?" *Proceedings of the 1990 ACM conference on Computer-supported cooperative work*, ACM, Los Angeles, USA, 357–370.

Matthews, O., and Howell, G. A. (2005). "Integrated project delivery an example of relational contracting." *Lean Construction Journal*, 2(1), 46–61.

Maturana, H. R., and Varela, F. J. (1987). *The tree of knowledge: The biological roots of human understanding*. New Science Library/Shambhala Publications, Boston, MA, US.

Maurer, M., and Lindemann, U. (2007). "Facing Multi-Domain Complexity in Product Development." *CiDaD Working Paper Series*, 03(01), 1 - 12.

Maurer, M. S. (2007). "Structural awareness in complex product design." Universität München, Verlag Dr. Hut, München, Germany.

McCord, K. R., and Eppinger, S. D. (1993). *Managing the integration problem in concurrent engineering*. Sloan School of Management Working Paper, Massachusetts Institute of Technology, Cambridge, USA.

McManus, H. (2005). *Product Development Value Stream Mapping (PDVSM) Manual 1.0*. Lean Aerospace Initiative, Massachusetts Institute of Technology, Cambridge, USA.

Mendling, J. (2008). *Metrics for Process Models: Empirical Foundations of Verification, Error Prediction, and Guidelines for Correctness.* Springer, München, Germany.

Meredith, J. (1998). "Building operations management theory through case and field research." *Journal of Operations Management*, 16(4), 441–454.

Mihm, J., Loch, C., and Huchzermeier, A. (2003). "Problem–Solving Oscillations in Complex Engineering Projects." *Management Science*, 49(6), 733–750.

Mohamad, A., Hickethier, G., Hovestadt, V., and Gehbauer, F. (2013). "Use of Modularization in Design as a Strategy to Reduce Component Variety in One-Off Projects." *Proceedings 21th Annual Conference of the International Group for Lean Construction (IGLC)*, Curitiba, Brazil.

Morelli, M. D., Eppinger, S. D., and Gulati, R. K. (1995) "Predicting technical communication in product development organizations." *IEEE Transactions on Engineering Management*, 42(3), 215–222.

Moreno, J. L. (1934). *Who shall survive?: A new approach to the problem of human interrelations.* Beacon House, Beacon, NY, USA.

Morgan, J. M., and Liker, J. K. (2006). *The Toyota Product Development System: Integrating People, Process, and Technology*. Productivity Press, New York, USA.

Mossman, A. (2015). "Last Planner®: 5 + 1 crucial & collaborative conversations for predictable design & construction delivery." <http://www.leanconstruction.org/media/docs/Mossman-Last-Planner> (May 24, 2015).

Motte, D., Bjärnemo, R., and Yannou, B. (2011). "On the integration of engineering design and development process model-Part I: Elaborations on the generally accepted process models." *3rd International Conference on Research into Design-ICoRD*, Bangalore, India, 87–95.

Newman, M. E. (2003). "The structure and function of complex networks." *SIAM review*, 45(2), 167–256.

Nonaka, I. (1990). "Redundant, overlapping organization: a Japanese approach to managing the innovation process." *California Management Review*, 32(3), 27–38.

O'Donnell, F. J., and Duffy, A. H. B. (2005). *Design performance*. Springer, London, UK.

Ohno, T. (1988). *Toyota production system: beyond large-scale production*. Productivity Press, New York, USA.

Oppenheim, B. W. (2004). "Lean product development flow." *Systems engineering*, 7(4), 352–376.

Padulo, L., and Arbib, M. A. (1974). *System Theory: A unified state-space approach to continuous and discrete systems*. Saunders, Philadelphia, USA.

Palla, G., Derényi, I., Farkas, I., and Vicsek, T. (2005). "Uncovering the overlapping community structure of complex networks in nature and society." *Nature*, 435(7043), 814–818.

Pall, G. A. (2000). *The process-centered enterprise: the power of commitments*. St. Lucie Press, New York, USA.

Parrish, K. D. (2009). "Applying a set-based design approach to reinforcing steel design." University of California, Berkeley, Berkeley, USA.

Pentland, A. (2012). "The new science of building great teams." *Harvard Business Review*, 90(4), 60–69.

Pessôa, M. V. P., Seering, W., Rebentisch, E., and Bauch, C. (2009). "Understanding the Waste Net: A Method for Waste Elimination Prioritization in Product Development." *Global Perspective for Competitive Enterprise, Economy and Ecology*, Advanced Concurrent Engineering, S.-Y. Chou, A. Trappey, J. Pokojski, and S. Smith, eds., Springer London, 233–242.

Pich, M. T., Loch, C. H., and De Meyer, A. (2002). "On uncertainty, ambiguity, and complexity in project management." *Management Science*, 1008–1023.

Piller, F. T., and Waringer, D. (1999). *Modularisierung in der Automobilindustrie: Neue Formen und Prinzipien; modular sourcing, Plattformkonzept und Fertigungssegmentierung als Mittel des Komplexitätsmanagements*. Shaker Verlag, Aachen, Germany.

Pimmler, T. U., and Eppinger, S. D. (1994). "Integration analysis of product decompositions." *ASME Design Theory and Methodology Conference*, Minneapolis, USA.

Priven, V., and Sacks, R. (2013). "Social Network Development in Last Planner System™ Implementations." *Proceedings 21th Annual Conference of the International Group for Lean Construction (IGLC)*, Fortaleza, Brazil.

Probst, G. (1987). *Selbst-Organisation: Ordnungsprozesse in sozialen Systemen aus ganzheitlicher Sicht*. P. Parey, Berlin, Germany.

Probst, G., and Büchel, B. (1997). *Organisationales Lernen: Wettbewerbsvorteil der Zukunft*. Dr. Th. Gabler Verlag, Wiesbaden, Germany.

Project Management Institute. (2008). *A guide to the project management body of knowledge (PMBOK Guide)*. Project Management Institute, Newtown Square, USA.

Pugh, D. S., Hickson, D. J., Hinings, C. R., and Turner, C. (1969). "The context of organization structures." *Administrative Science Quarterly*, 91–114.

Pugh, S. (1991). *Total design: integrated methods for successful product engineering*. Addison-Wesley, Wokingham, UK.

Pulm, U. (2004). *Eine systemtheoretische Betrachtung der Produktentwicklung*. Verlag Dr. Hut, München, Germany.

Reichardt, T., Elezi, F., Tommelein, I. D., and Lindemann, U. (2012). "Creating dynamic organizational modularity in Lean Construction Design — Combining MDM and DSM Methodology systematically." *Gain competitive advantage by managing complexity: Proceedings of the 14th International DSM Conference*, Kyoto, Japan.

Reichwald, R., and Möslein, K. (1997). *Organisation: Strukturen und Gestaltung*. Arbeitsberichte des Lehrstuhls für Allgemeine und Industrielle Betriebswirtschaftslehre an der Technischen Universität München, Techn. Univ. München, München, Germany.

Reinertsen, D. G. (1997). *Managing the design factory: a product developer's toolkit*. The Free Press, New York.

Remington, K., Zolin, R., and Turner, R. (2009). "A model of project complexity : distinguishing dimensions of complexity from severity." *Proceedings of the 9th International Research Network of Project Management Conference*, Berlin, Germany.

Rittel, H. W. J., and Webber, M. M. (1973). "Dilemmas in a general theory of planning." *Policy sciences*, 4(2), 155–169.

Rother, M. (2009). *Toyota Kata: managing people for improvement, adaptiveness and superior results*. McGraw-Hill Professional, New York, USA.

Rother, M., and Shook, J. (2003). *Learning to See: Value Stream Mapping to Add Value and Eliminate Muda*. Lean Enterprise Institute, Cambridge, USA.

Russell, M. M. (2013). "Allocation of Time Buffer to Construction Project Task Durations." North Carolina State University, Raleigh, USA.

Sauser, B. J., Reilly, R. R., and Shenhar, A. J. (2009). "Why projects fail? How contingency theory can provide new insights – A comparative analysis of NASA's Mars Climate Orbiter loss." *International Journal of Project Management*, 27(7), 665–679.

Schindler, M., and Eppler, M. J. (2003). "Harvesting project knowledge: a review of project learning methods and success factors." *International Journal of Project Management*, 21(3), 219–228.

Schlindwein, S. L., and Ison, R. (2004). "Human knowing and perceived complexity: Implications for systems practice." *Emergence: Complexity & Organization*, 6(3), 27–32.

Schöttle, A., and Gehbauer, F. (2013). "Incentive Structure in Public Design-Bid-Build Tendering and its Effects on Projects." *Proceedings of the 21th Annual Conference of the International Group for Lean Construction (IGLC 21)*, Fortaleza, Brazil.

Schrader, S., Riggs, W. M., and Smith, R. P. (1993). "Choice over uncertainty and ambiguity in technical problem solving." *Journal of Engineering and Technology Management*, 10(1-2), 73–99.

Shah, R., and Ward, P. T. (2007). "Defining and developing measures of lean production." *Journal of operations management*, 25(4), 785–805.

Shannon, C. E., and Weaver, W. (1959). *The mathematical theory of communication*. University of Illinois Press, Champaign, USA.

Sheard, S. (2007). "Complex adaptive systems in systems engineering and management." *Handbook of Systems Engineering and Management*, A. P. Sage and W. B. Rouse, eds., Wiley, New York, USA, 1283–1318.

Sheffer, D. A. (2011). "Innovation in modular industries: Implementing energy-efficient innovations in US buildings." Stanford University, Stanford, USA.

Shenhar, A. J., and Dvir, D. (1996). "Toward a typological theory of project management." *Research policy*, 25(4), 607–632.

Sherman, J. D. (2004). "Optimal modes and levels of integration, and the identification of cross-functional coordination deficiencies in concurrent engineering." *IEEE Transactions on Engineering Management*, 51(3), 268–278.

Shewhart, W. A. (1939). *Statistical method from the viewpoint of quality control*. US Department of Agriculture, Washington, USA.

Shingo, S. (1989). *A Study of the Toyota Production System: From an Industrial Engineering Viewpoint*. Productivity Press, New York, USA.

Simon, H. A. (1962). "The architecture of complexity." *Proceedings of the american philosophical society*, 467–482.

Simon, H. A. (1996). *The sciences of the artificial*. MIT Press, Cambridge, MA.

Smith, R. P., and Eppinger, S. D. (1997). "A predictive model of sequential iteration in engineering design." *Management Science*, 43(8), 1104–1120.

Sosa, M. E. (2008). "A structured approach to predicting and managing technical interactions in software development." *Research in Engineering Design*, 19(1), 47–70.

Sosa, M. E., Eppinger, S. D., Pich, M., McKendrick, D. G., and Stout, S. K. (2002). "Factors that influence technical communication in distributed product development: an empirical study in the telecommunications industry." *IEEE Transactions on Engineering Management*, 49(1), 45–58.

Sosa, M. E., Eppinger, S. D., and Rowles, C. M. (2004). "The misalignment of product architecture and organizational structure in complex product development." *Management Science*, 1674–1689.

Sosa, M. E., Eppinger, S. D., and Rowles, C. M. (2007). "Are your engineers talking to one another when they should?" *Harvard Business Review*, 85(11), 133.

Sosa, M., and Mihm, J. (2008). "Organization design for new product development." *Handbook of New Product Development Management*, C. H. Loch and S. Kavadias, eds., Taylor & Francis, New York, USA, 165–198.

Sparapani, A. (2011). "Van Ness and Geary Hospital Project Delivery Guide v.3.2." Van Ness and Geary Hospital Project, San Francisco, USA.

Spear, S., and Bowen, H. K. (1999). "Decoding the DNA of the Toyota production system." *Harvard Business Review*, 77, 96–108.

Stachowiak, H. (1973). *Allgemeine Modelltheorie*. Springer-Verlag, Wien, Austria.

Steward, D. V. (1962). "On an Approach to Techniques for the Analysis of the Structure of Large Systems of Equations." *SIAM Review*, 4(4), 321–342.

Steward, D. V. (1981). "The design structure system: A method for managing the design of complex systems." *IEEE Transactions on Engineering Management*, (3), 71–74.

Thompson, J. (2010). *Organizations in Action: Social Science Bases of Administrative Theory. Classics in Organization and Management Series*. Trans. Publishers, New Brunswick, USA.

Thomsen, C., Darrington, J., Dunne, D., and Lichtig, W. (2010). *Managing integrated project delivery*. White paper, Construction Management Association of America, McLean, USA.

Tommelein, I. D., and Gholami, S. (2012). "Root Causes of Clashes in Building Information Models." *Proceedings 20th Annual Conference of the International Group for Lean Construction (IGLC)*, San Diego, CA.

Tribelsky, E., and Sacks, R. (2010). "The Relationship between Information Flow and Project Success in Multi-Disciplinary Civil Engineering Design." *Proceedings 18th Annual Conference of the International Group for Lean Construction (IGLC)*, Haifa, Israel.

Tsao, C. C., Tommelein, I. D., Swanlund, E., and Howell, G. A. (2000). "Case study for work structuring: Installation of metal door frames." *Proceedings 8th Annual Conference of the International Group for Lean Construction (IGLC)*, Brigthon, UK.

Tsao, C., Tommelein, I., Swanlund, E., and Howell, G. (2004). "Work Structuring to Achieve Integrated Product–Process Design." *Journal of Construction Engineering and Management*, 130(6), 780–789.

Tuholski, S. J. (2008). "Transformation, Flow, and Value Constellations in AEC Projects." University of California, Berkeley, Berkeley, USA.

Tuholski, S. J., and Tommelein, I. D. (2008). "Design Structure Matrix (DSM) Implementation on a Seismic Retrofit." *Proceedings 16th Annual Conference of the International Group for Lean Construction (IGLC)*, Manchester, UK.

Tushman, M. L., and Nadler, D. A. (1978). "Information Processing as an Integrating Concept in Organizational Design." *Academy of Management Review*, 3(3), 613–624.

Ulrich, K. T., and Eppinger, S. D. (2004). *Product design and development*. McGraw-Hill/Irwin, New York, USA.

Vajna, S. (2005). "Workflow for design." *Design process improvement - a review of current practise*, J. Clarkson and C. Eckert, eds., Springer, London, UK, 366–385.

Veenstra, V. S., Halman, J. I., and Voordijk, J. T. (2006). "A methodology for developing product platforms in the specific setting of the housebuilding industry." *Research in engineering design*, 17(3), 157–173.

Vester, F. (2002). *Die Kunst vernetzt zu denken*. dtv Wissen, München, Germany.

Vidal, L.-A., and Marle, F. (2008). "Understanding project complexity: implications on project management." *Kybernetes*, 37(8), 1094–1110.

Waldron, M. B., and Waldron, K. J. (1996). "Design characterizations." *Mechanical design: Theory and methodology*, M. B. Waldron and K. J. Waldron, eds., Springer, New York, USA, 35–51.

Ward, A., Liker, J. K., Cristiano, J. J., and Sobek II, D. K. (1995). "The Second Toyota Paradox: How Delaying Decisions Can Make Better Cars Faster." *Sloan Management Review*, (36), 43–61.

Wasserman, S., and Faust, K. (1994). *Social Network Analysis: Methods and Applications*. Cambridge University Press, Cambridge, USA.

Wasson, C. S. (2006). *System Analysis, Design, and Development: Concepts, Principles, and Practices*. John Wiley & Sons, Hoboken, USA.

Watts, D. J., and Strogatz, S. H. (1998). "Collective dynamics of 'small-world' networks." *Nature*, 393(6684), 440–442.

Weaver, W. (1948). "Science and complexity." *American scientist*, 36(4), 536–544.

De Weck, O. L. (2007). "On the role of DSM in designing systems and products for changeability." *Proceedings of 9th International DSM Conference*, München, Germany, 311–324.

Williams, T. (2005). "Assessing and moving on from the dominant project management discourse in the light of project overruns." *IEEE Transactions on Engineering Management*, 52(4), 497–508.

Williams, T. M. (1999). "The need for new paradigms for complex projects." *International Journal of Project Management*, 17(5), 269–273.

Womack, J. P., and Jones, D. T. (2010). *Lean thinking: banish waste and create wealth in your corporation*. Simon and Schuster, New York, USA.

Womack, J. P., Jones, D. T., and Roos, D. (1990). *The Machine that Changed the World: The Story of Lean Production.* Rawson Associates, New York, USA.

Worren, N. A. (2012). *Organisation design: Re-defining complex systems.* Pearson, London, UK.

Wynn, D. C., Eckert, C. M., and Clarkson, P. J. (2007). "Modelling iteration in engineering design." *16th International Conference on Engineering Design (ICED)*, Paris, France.

Wynn, D., and Clarkson, J. (2005). "Models of designing." *Design process improvement - a review of currrent practise*, J. Clarkson and C. Eckert, eds., Springer, New York, USA, 34–59.

Yassine, A., Whitney, D., Daleiden, S., and Lavine, J. (2003). "Connectivity maps: Modeling and analysing relationships in product development processes." *Journal of Engineering Design*, 14(3), 377–394.

Yin, R. K. (2009). *Case Study Research: Design and Methods*. SAGE, London, UK.

Zamenopoulos, T., and Alexiou, K. (2005). "Linking design and complexity: a review." *Proceedings of the ECCS 2005 Satellite Workshop: Embracing Complexity in Design*, Paris, France.

Zimina, D., Ballard, G., and Pasquire, C. (2012). "Target value design: using collaboration and a lean approach to reduce construction cost." *Construction Management and Economics*, 30(5), 383–398.

Appendices

Appendix A: Information Exchange between People in Case Study A

Figure 70: Information Exchange between People sorted by Company

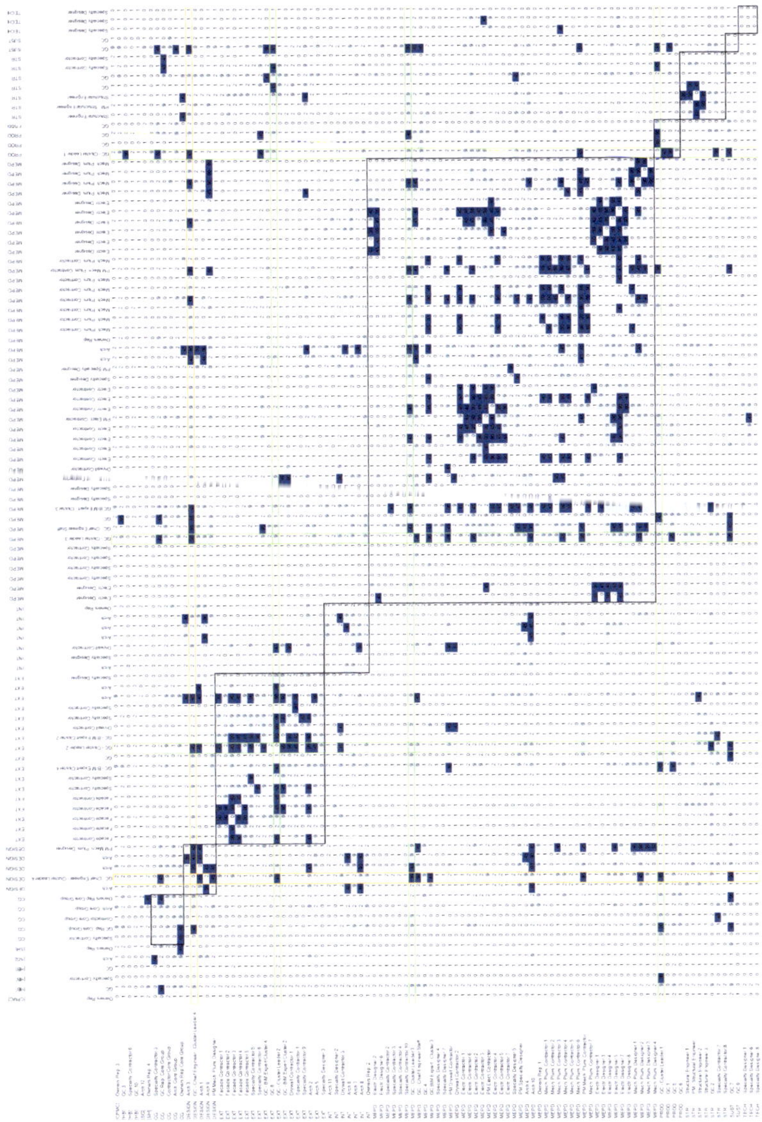

Figure 71: Information Exchange between People sorted by Cluster Group with marked Chief Engineer (yellow) and Cluster Leaders (green)

Appendix B: Sensitivity Analysis for Case Study A

This part of the appendix presents the sensitivity analysis of the case study "The project as a Virtual Company". Sensitivity analysis compared three different social networks, all which are derived from the same dataset. This dataset consists of 99 people and contains their communication from two perspectives: perspective one indicates the information a person 'gives' to other people, perspective two indicates the information a person 'receives' from other people.

Three different social networks were derived from the initial data, and these three different social networks are based on three different combinations of the give and receive perspective:

1. Max-function:

$$\text{information exchange (Person A, Person B)} \\ = \max\left[\text{receive (Person A, Person B), give (Person B, Person A)}\right]$$

2. Min-function:

$$\text{information exchange (Person A, Person B)} \\ = \min\left[\text{receive (Person A, Person B), give (Person B, Person A)}\right]$$

1. Mean-function:

$$\text{information exchange (Person A, Person B)} \\ = \frac{\text{receive (Person A, Person B)} + \text{give (Person B, Person A)}}{2}$$

7.4.1 1. Visual Comparison of Force-directed Graphs

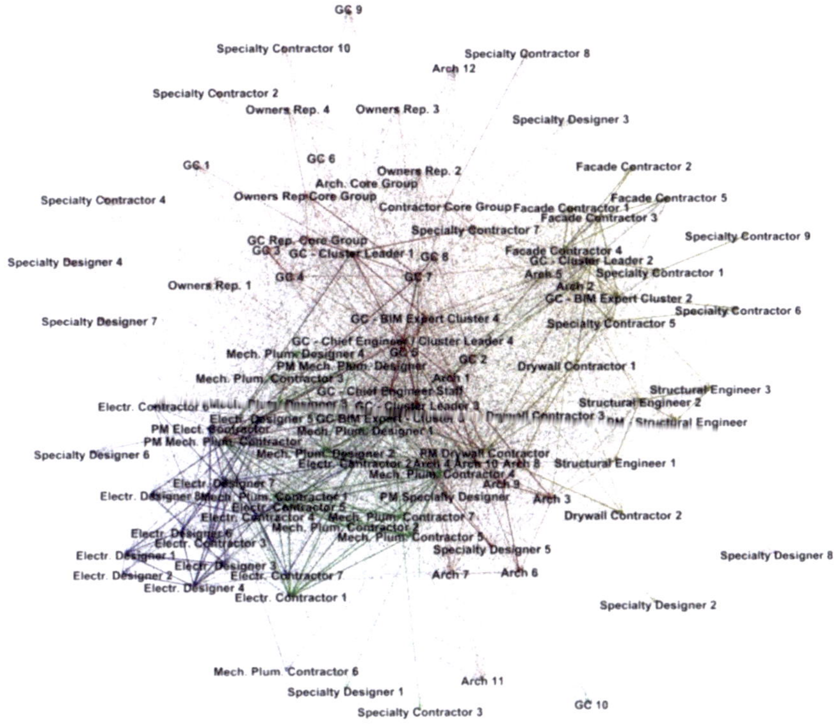

Figure 72: Force-directed Graph of Communication at VNGC based on Max-Values, all Levels of Communication

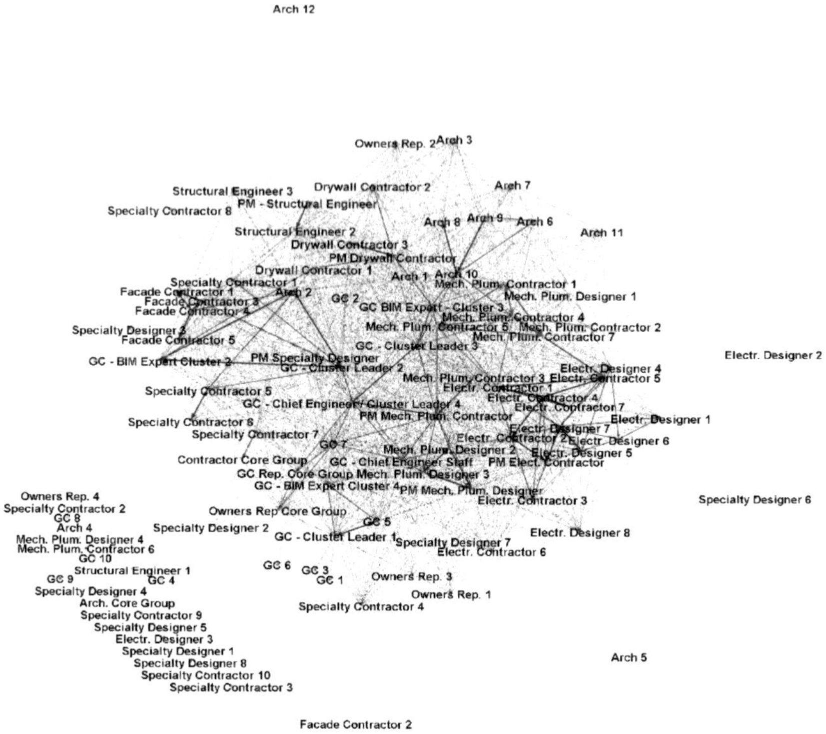

Figure 73: Force-directed Graph of Communication at VNGC based on Min-Values, all Levels of Communication

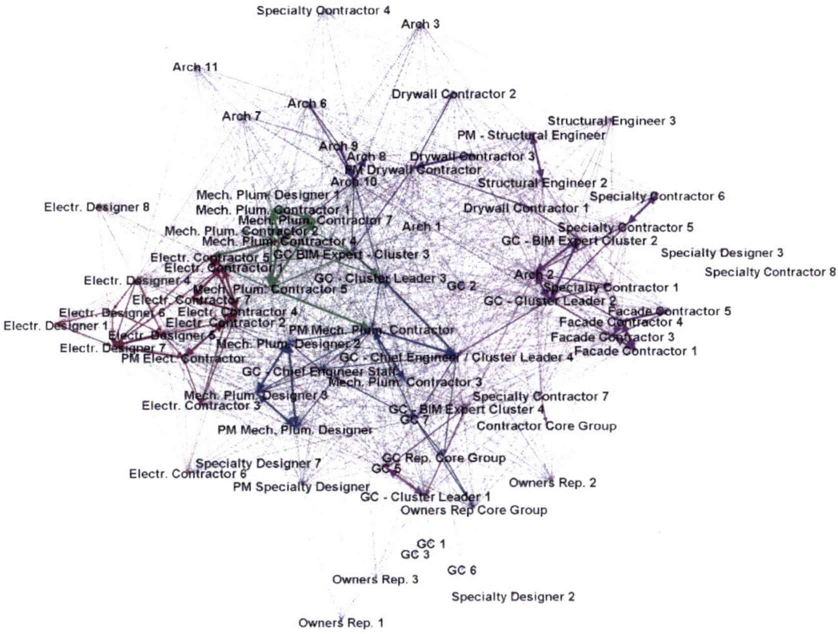

Figure 74: Force-directed Graph of Communication at VNGC based on Min-Values, without disconnected Nodes, all Levels of Communication

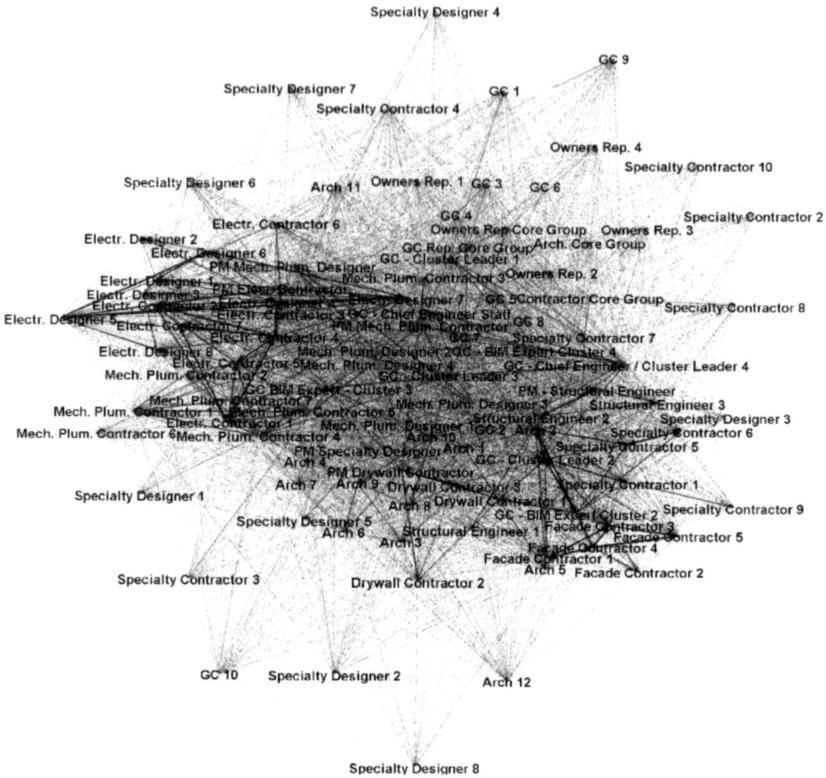

Figure 75: Force-directed Graph of Communication at VNGC based on Mean-Values, all Levels of Communication

7.4.2 2. Comparison of Centralities

The following tables 32, 33, and 34 present respectively degree centralities, betweenness centralities, and closeness centralities of people in the SNA model. Tables 32, 33, and 34 show the 15 highest ranking people for each type of centrality in descending order. Numbers following roles, e.g., Electr. Designer, stem from anonymizing the data-set und represent a sequential numbering of people with the same role.

Table 32: Comparison of Degree Centralities for three Scenarios

Max-Value	**Min-Value**	**Mean-Value**
PM Mech. Plum. Contractor	Mech. Plum. Contractor 4	Mech. Plum. Contractor 4
Mech. Plum. Contractor 4	PM Mech. Plum. Contractor	PM Mech. Plum. Contractor
Electr. Designer 7	Electr. Contractor 1	Electr. Designer 7
GC - Cluster Leader 3	Electr. Contractor 4	GC - Chief Engineer / Cluster Leader 4
GC - Chief Engineer / Cluster Leader 4	Mech. Plum. Contractor 5	GC - Cluster Leader 3
GC BIM Expert - Cluster 3	GC - Cluster Leader 2	GC BIM Expert - Cluster 3
GC - Cluster Leader 2	GC - Cluster Leader 3	GC - Cluster Leader 2
Arch 10	Electr. Contractor 2	Electr. Contractor 1
Electr. Contractor 2	Electr. Designer 7	Electr. Contractor 4
Arch 2	GC - Chief Engineer / Cluster Leader 4	Electr. Contractor 2
Electr. Contractor 4	Arch 2	Arch 10
Mech. Plum. Designer 2	GC BIM Expert - Cluster 3	Arch 2
Electr. Contractor 5	Arch 10	Electr. Contractor 5
Electr. Contractor 1	PM Elect. Contractor	Mech. Plum. Contractor 5
Mech. Plum. Designer 2	Mech. Plum. Contractor 1	Mech. Plum. Designer 2

Table 33: Comparison of Betweenness Centralities for three Scenarios

Max-Value	Min-Value	Mean-Value
GC - Chief Engineer / Cluster Leader 4	GC - Cluster Leader 1	GC - Chief Engineer / Cluster Leader 4
Owners Rep Core Group	Owners Rep Core Group	Owners Rep Core Group
GC - Chief Engineer Staff	GC - Chief Engineer Staff	GC - Chief Engineer Staff
GC - Cluster Leader 1	Electr. Designer 5	GC - Cluster Leader 1
GC - Cluster Leader 3	Specialty Contractor 7	GC - Cluster Leader 3
GC - BIM Expert Cluster 4	GC - Cluster Leader 3	GC - BIM Expert Cluster 4
PM Mech. Plum. Contractor	Mech. Plum. Contractor 4	PM Mech. Plum. Contractor
GC 7	Arch 10	GC 7
GC - Cluster Leader 2	PM Mech. Plum. Contractor	GC - Cluster Leader 2
GC 5	GC - BIM Expert Cluster 4	GC 5
Structural Engineer 3	GC - Cluster Leader 2	Structural Engineer 3
Drywall Contractor 3	PM Drywall Contractor	Drywall Contractor 3
Structural Engineer 2	Arch 2	Structural Engineer 2
PM Drywall Contractor	GC - Chief Engineer / Cluster Leader 4	PM Drywall Contractor
Owners Rep. 2	GC 7	Owners Rep. 2

Table 34: Comparison of Closeness Centralities for three Scenarios

Max-Value	**Min-Value**	**Mean-Value**
GC 6	Contractor Core Group	GC 6
Contractor Core Group	GC Rep. Core Group	Contractor Core Group
Specialty Designer 8	GC 3	Specialty Designer 8
Specialty Contractor 9	GC 1	Specialty Contractor 9
Specialty Contractor 10	GC 6	Specialty Contractor 10
Specialty Contractor 2	Mech. Plum. Contractor 2	Specialty Contractor 2
Specialty Designer 4	Electr. Designer 6	Specialty Designer 4
Specialty Designer 7	Specialty Designer 7	Specialty Designer 7
Specialty Designer 3	Specialty Designer 3	Specialty Designer 3
Specialty Contractor 3	Specialty Designer 2	Specialty Contractor 3
Electr. Contractor 7	Electr. Contractor 7	Electr. Contractor 7
Specialty Designer 6	Arch 1	Specialty Designer 6
Specialty Designer 5	Specialty Contractor 8	Specialty Designer 5
Electr. Designer 3	Facade Contractor 4	Electr. Designer 3
Electr. Designer 2	Mech. Plum. Contractor 7	Electr. Designer 2

FORSCHUNGSBERICHTE DES INSTITUTS FÜR TECHNOLOGIE UND MANAGEMENT IM BAUBETRIEB

Heft 1–62 institutsintern verlegt

Heft 1 **HANS PINNOW** 1972
Vergleichende Untersuchungen von Tiefbauprojekten in offener Bauweise

Heft 2 **HEINRICH MÜLLER** 1972
Rationalisierung des Stahlbetonbaus durch neue Schalverfahren und deren Optimierung beim Entwurf

Heft 3 **DIETER KARLE** 1972
Einsatzdimensionierung langsam schlagender Rammbäre aufgrund von Rammsondierungen

Heft 4 **WILHELM REISMANN** 1973
Kostenerfassung im maschinellen Erdbau

Heft 5 **GÜNTHER MALETON** 1973
Wechselwirkungen von Maschine und Fels beim Reißvorgang

Heft 6 **JOACHIM HORNUNG** 1973
Verfahrenstechnische Analyse über den Ersatz schlagender Rammen durch die Anwendung lärmarmer Baumethoden

Heft 7 **THOMAS TRÜMPER / JÜRGEN WEID** 1973
Untersuchungen zur optimalen Gestaltung von Schneidköpfen bei Unterwasserbaggerungen

Heft 8 **GEORG OELRICHS** 1974
Die Vibrationsrammung mit einfacher Längsschwingwirkung – Untersuchungen über die Kraft- und Bewegungsgrößen des Systems Rammbär plus Rammstück im Boden

Heft 9 **PETER BÖHMER** 1974
Verdichtung bituminösen Mischgutes beim Einbau mit Fertigern

Heft 10 **FRITZ GEHBAUER** 1974
Stochastische Einflußgrößen für Transportsimulationen im Erdbau

Heft 11	**EMIL MASSINGER** Das rheologische Verhalten von lockeren Erdstoffgemischen	1976
Heft 12	**KAWUS SCHAYEGAN** Einfluß von Bodenkonsistenz und Reifeninnendruck auf die fahrdynamischen Grundwerte von EM-Reifen	1975
Heft 13	**CURT HEUMANN** Dynamische Einflüsse bei der Schnittkraftbestimmung in standfesten Böden	1975
Heft 14	**HANS-JOSEF KRÄMER** Untersuchung der bearbeitungstechnischen Bodenkennwerte mit schwerem Ramm-Druck-Sondiergerät zur Beurteilung des Maschineneinsatzes im Erdbau	1976
Heft 15	**FRIEDRICH ULBRICHT** Baggerkraft bei Eimerkettenschwimmbaggern - Untersuchungen zur Einsatzdimensionierung	1977
Heft 16	**BERTOLD KETTERER** Einfluß der Geschwindigkeit auf den Schneidvorgang in rolligen Böden **- vergriffen -**	1977
Heft 17	**JOACHIM HORNUNG/THOMAS TRÜMPER** Entwicklungstendenzen lärmarmer Tiefbauverfahren für den innerstädtischen Einsatz	1977
Heft 18	**JOACHIM HORNUNG** Geometrisch bedingte Einflüsse auf den Vorgang des maschinellen Reißens von Fels - untersucht an Modellen	1978
Heft 19	**THOMAS TRÜMPER** Einsatzoptimierung von Tunnelvortriebsmaschinen	1978
Heft 20	**GÜNTHER GUTH** Optimierung von Bauverfahren - dargestellt an Beispielen aus dem Seehafenbau	1978
Heft 21	**KLAUS LAUFER** Gesetzmäßigkeiten in der Mechanik des drehenden Bohrens im Grenzbereich zwischen Locker- und Festgestein **- vergriffen -**	1978

Heft 22	**URS BRUNNER** Submarines Bauen - Entwicklung eines Bausystems für den Einsatz auf dem Meeresboden **- vergriffen -**	1979
Heft 23	**VOLKER SCHULER** Drehendes Bohren in Lockergestein - Gesetzmäßigkeiten und Nutzanwendung **- vergriffen -**	1979
Heft 24	**CHRISTIAN BENOIT** Die Systemtechnik der Unterwasserbaustelle im Offshore-Bereich	1980
Heft 25	**BERNHARD WÜST** Verbesserung der Umweltfreundlichkeit von Maschinen, insbesondere von Baumaschinen-Antrieben	1980
Heft 26	**HANS-JOSEF KRÄMER** Geräteseitige Einflußparameter bei Ramm- und Drucksondierungen und ihre Auswirkungen auf den Eindringwiderstand	1981
Heft 27	**BERTOLD KETTERER** Modelluntersuchungen zur Prognose von Schneid- und Planierkräften im Erdbau	1981
Heft 28	**HARALD BEITZEL** Gesetzmäßigkeiten zur Optimierung von Betonmischern	1981
Heft 29	**BERNHARD WÜST** Einfluß der Baustellenarbeit auf die Lebensdauer von Turmdrehkranen	1982
Heft 30	**HANS PINNOW** Einsatz großer Baumaschinen und bisher nicht erfaßter Sonderbauformen in lärmempfindlichen Gebieten	1982
Heft 31	**WALTER BAUMGÄRTNER** Traktionsoptimierung von EM-Reifen in Abhängigkeit von Profilierung und Innendruck	1982
Heft 32	**KARLHEINZ HILLENBRAND** Wechselwirkung zwischen Beton und Vibration bei der Herstellung von Stahlbetonrohren im Gleitverfahren	1983
Heft 33	**CHRISTIAN BENOIT** Ermittlung der Antriebsleistung bei Unterwasserschaufelrädern	1985

Heft 34	**NORBERT WARDECKI** Strömungsverhalten im Boden-/Werkzeugsystem	1986
Heft 35	**CHRISTIAN BENOIT** Meeresbergbau - Bestimmung der erforderlichen Antriebskraft von Unterwasserbaggern	1986
Heft 36	**ROLF VICTOR SCHMÖGER** Automatisierung des Füllvorgangs bei Scrapern	1987
Heft 37	**ALEXANDER L. MAY** Analyse der dreidimensionalen Schnittverhältnissen beim Schaufelradbagger	1987
Heft 38	**MICHAEL HELD** Hubschraubereinsatz im Baubetrieb	1989
Heft 39	**GUNTER SCHLICK** Adhäsion im Boden-Werkzeug-System	1989
Heft 40	**FRANZ SAUTER** Optimierungskriterien für das Unterwasser- schaufelrad (UWS) mittels Modellsimulation **- vergriffen -**	1991
Heft 41	**STEFAN BERETITSCH** Kräftespiel im System Schneidwerkzeug-Boden	1992
Heft 42	**HEINRICH SCHLICK** Belastungs- und Fließverhältnisse in Silos mit zentralen Einbauten und Räumarmaustrag	1994
Heft 43	**GÜNTHER DÖRFLER** Untersuchungen der Fahrwerkbodeninteraktion zur Gestaltung von Raupenfahrzeugen für die Befahrung weicher Tiefseeböden	1995
Heft 44	**AXEL OLEFF** Auslegung von Stellelementen für Schwingungserregerzellen mit geregelter Parameterverstellung und adaptive Regelungskonzepte für den Vibrationsrammprozeß	1996
Heft 45	**KUNIBERT LENNERTS** Stand der Forschung auf den Gebieten der Facility- und Baustellen-Layoutplanung	1997
Heft 46	**KUNIBERT LENNERTS** Ein hybrides, objektorientiertes System zur Planung optimierter Baustellen-Layouts	1997

Heft 47 **UWE RICKERS** 1998
Modellbasiertes Ressourcenmanagement
für die Rettungsphase in Erdbebengebieten

Heft 48 **ULRICH-PETER REHM** 1998
„Ermittlung des Antriebsdrehmomentes von
Räumarmen in Silos mit Einbaukörper und
kohäsivem Schüttgut"

Heft 49 **DIRK REUSCH** 2001
Modellierung, Parameterschätzung und
automatische Regelung mit Erschütterungsbegrenzung
für das langsame Vibrationsrammen

Heft 50 **FRANZ DIEMAND** 2001
Strategisches und operatives Controlling
im Bauunternehmen

Heft 51 **KARSTEN SCHÖNBERGER** 2002
Entwicklung eines Workflow-Management-
Systems zur Steuerung von Bauprozessen in
Handwerkernetzwerken

Heft 52 **CHRISTIAN MEYSENBURG** 2002
Ermittlung von Grundlagen für das Controlling
in öffentlichen Bauverwaltungen

Heft 53 **MATTHIAS BURCHARD** 2002
Grundlagen der Wettbewerbsvorteile globaler
Baumärkte und Entwicklung eines Marketing
Decision Support Systems (MDSS) zur
Unternehmensplanung

Heft 54 **JAROSŁAW JURASZ** 2003
Geometric Modelling for Computer Integrated Road
Construction (Geometrische Modellierung für den
rechnerintegrierten Straßenbau)

Heft 55 **SASCHA GENTES** 2003
Optimierung von Standardbaumaschinen
zur Rettung Verschütteter

Heft 56 **GERHARD W. SCHMIDT** 2003
Informationsmanagement und
Transformationsaufwand im Gebäudemanagement

Heft 57 **KARL LUDWIG KLEY** 2004
Positionierungslösung für Straßenwalzen -
Grundlage für eine kontinuierliche Qualitätskontrolle
und Dokumentation der Verdichtungsarbeit im Asphaltbau

Heft 58	**JOCHEN WENDEBAUM** Nutzung der Kerntemperaturvorhersage zur Verdichtung von Asphaltmischgut im Straßenbau	2004
Heft 59	**FRANK FIEDRICH** Ein High-Level-Architecture-basiertes Multiagentensystem zur Ressourcenoptimierung nach Starkbeben	2004
Heft 60	**JOACHIM DEDEKE** Rechnergestützte Simulation von Bauproduktions-prozessen zur Optimierung, Bewertung und Steuerung von Bauplanung und Bauausführung	2005
Heft 61	**MICHAEL OTT** Fertigungssystem Baustelle - Ein Kennzahlensystem zur Analyse und Bewertung der Produktivität von Prozessen	2007
Heft 62	**JOCHEN ABEL** Ein produktorientiertes Verrechnungssystem für Leistungen des Facility Management im Krankenhaus	2007

HEFT 63–68 BEI KIT SCIENTIFIC PUBLISHING KARLSRUHE VERLEGT, ISSN 1868-5951

Heft 63	**JÜRGEN KIRSCH** Organisation der Bauproduktion nach dem Vorbild industrieller Produktionssysteme – Entwicklung eines Gestaltungsmodells eines ganzheitlichen Produktionssystems für den Bauunternehmer	2009
Heft 64	**MARCO ZEIHER** Ein Entscheidungsunterstützungsmodell für den Rückbau massiver Betonstrukturen in kerntechnischen Anlagen	2009
Heft 65	**MARKUS SCHÖNIT** Online-Abschätzung der Rammguttragfähigkeit beim langsamen Vibrationsrammen in nichtbindigen Böden	2009
Heft 66	**JOHANNES KARL WESTERMANN** Betonbearbeitung mit hydraulischen Anbaufräsen	2009
Heft 67	**FABIAN KOHLBECKER** Projektbegleitendes Öko-Controlling Ein Beitrag zur ausgewogenen Bauprojektrealisierung beispielhaft dargestellt anhand von Tunnelbauprojekten	2010
Heft 68	**AILKE HEIDEMANN** Kooperative Projektabwicklung im Bauwesen unter der Berücksichtigung von Lean-Prinzipien - Entwicklung eines Lean-Projektabwicklungssystems: Internationale Untersuchungen im Hinblick auf die Umsetzung und Anwendbarkeit in Deutschland	2011

AB HEFT 69 BEI KIT SCIENTIFIC PUBLISHING KARLSRUHE UNTER DEM TITEL KARLSRUHER REIHE TECHNOLOGIE UND MANAGEMENT IM BAUBETRIEB VERLEGT, ISSN 2363-8222

Heft 69	**KIM KIRCHBACH** Anwendung von Lean-Prinzipien im Erdbau – Entwicklung eines Baustellenleitstands auf Basis von Virtual Reality	2015
Heft 70	**PATRICK KERN** Elastomerreibung und Kraftübertragung beim Abscheren von aktiv betriebenen Vakuumgreifern auf rauen Oberflächen	2017
Heft 71	**GERNOT HICKETHIER** Communication Structures in the Design Phase of Lean Project Delivery	2019

Die Bände sind unter www.ksp.kit.edu als PDF frei verfügbar oder als Druckausgabe bestellbar.